Animal Health in China

(2014)

Veterinary Bureau
Ministry of Agriculture, P. R. China

CHINA AGRICULTURE PRESS

图书在版编目（CIP）数据

中国动物卫生状况报告. 2014：英文 / 农业部兽医局编. — 北京：中国农业出版社，2016.1
ISBN 978-7-109-21660-0

Ⅰ. ①中… Ⅱ. ①农… Ⅲ. ①家畜卫生－研究报告－中国－2014－英文 Ⅳ. ①S851.2

中国版本图书馆CIP数据核字(2016)第108656号

中国农业出版社出版
（北京市朝阳区麦子店街18号楼）
（邮政编码100125）
责任编辑　邱利伟　周锦玉

北京通州皇家印刷厂印刷　新华书店北京发行所发行
2016年1月第1版　2016年1月北京第1次印刷

开本：889mm×1194mm　1/16　印张：7.5
字数：150千字
定价：68.00元
（凡本版图书出现印刷、装订错误，请向出版社发行部调换）

Preface

The year of 2014 was the first year for China to deepen reforms comprehensively. To prevent serious regional animal epidemics and safety quality incidents of agricultural products, veterinary departments at all levels adopted method of performance management and extended it to intensify the prevention and control of the major animal epidemics. The veterinary authority insisted on reform and innovation, governed in the strict accordance with the law, scientifically studied and judged the developmental situation and explored new work methods and management modes as well as measures actively. As the legal system was being gradually improved, we took great efforts to strengthen the prevention and control of animal epidemics, especially the key steps in immunization, surveillance, and emergency disposal. Animal health status has been stable and steady with no regional major animal epidemics in 2014. Although avian influenza and foot-and-mouth disease were reported, these epidemics were eliminated immediately. Besides, we not only reacted orderly to the H7N9 influenza, but also controlled and eliminated the peste des petits ruminants (PPR) epidemic that re-emerged in China. In addition, we successfully dealt with emergencies such as flood, debris flow, landslide and super typhoon, securing the safety of animal farming, animal-origin food safety and public health. Zoning and compartmentalization was pushed forward and construction of Big Northeast Zone free FMD without vaccination was launched. Through deepening reform on veterinary system, we moved forward the adjustment of domestic slaughter regulating function, improved the supervision on slaughter industry, perfected supervision system of livestock slaughter statistics and implemented accountability for quality safety of slaughter products. In order to intensify veterinary law-enforcement work, the authority launched the action of standard improvement and ability strengthen for animal health supervisors all around China. By establishing long-term mechanism for bio-safety disposal of animals died of illness, the Bureau also enhanced the monitoring of animal bodies mentioned above. More restrict quality supervision on inputs and veterinary drugs such as vaccines for major animal disease has been initiated when rectification, inspection and sample, test and penalty activities on veterinary drugs were carried out. The monitoring plan of veterinary drug residue was adopted in all aspects as well as Antibiotics rectification.QR-code-traceback pilot project for veterinary drug was initiated, strictly suppressing illegal drug abuse so quality and safety level of animal products are further improved. Official veterinary qualification verification continues, as well as national licensed veterinarian qualification examination and licensed veterinary qualification conferring. Team development is fortified for official veterinaries, licensed veterinaries and grassroots level veterinaries. Cooperation and exchange are further deepened and promoted both bilaterally and multilaterally, with international organizations, countries, Hong Kong, Macau and Taiwan regions. The across-boundary joint epidemic control mechanism is further fortified with neighboring countries. International cooperation has gained fruitful achievements in 2014.

2015 is the last year of the twelfth 5-year plan, during which the job of veterinary becomes more arduous and onerous. Veterinarians at all levels should further emancipate our mind, coordinate our work, push forward reform and innovation and then earnestly bring diverse measures into force. Only in that way, we can open a new situation for veterinary work, contributing more to the development of China's veterinary course.

Director General,
Veterinary Bureau,
Ministry of Agriculture, P. R. China

In January 2015

Contents

Preface

Chapter 1 Veterinary System ·· 1

1.1 Veterinary agencies and organizations ··· 1

    1.1.1 Veterinary administrative agencies ·· 1

    1.1.2 Veterinary administrative law enforcement agencies ·················· 3

    1.1.3 Veterinary technical support system ······································ 3

    1.1.4 Animal production management and technology promotion institutions ·· 5

    1.1.5 Entry-exit inspection and quarantine system ··························· 6

    1.1.6 Veterinary scientific research system ····································· 6

    1.1.7 Veterinary higher education system ······································ 9

    1.1.8 Societies, associations and professional technical committees ······ 9

1.2 Capacity building of veterinarians ·· 13

    1.2.1 Official veterinary ··· 13

    1.2.2 licensed veterinarians ·· 13

    1.2.3 Rural veterinarian ··· 14

1.3 PVS evaluation ·· 14

Chapter 2 Veterinary Laws and Regulations ········· 16

Chapter 3 Prevention and Control of Animal Disease ········· 28

- 3.1 OIE listed diseases that were never reported or already eradicated in China ········· 28
- 3.2 Prevention and control of major animal diseases ········· 30
  - 3.2.1 Foot-and-mouth disease ········· 30
  - 3.2.2 HPAI ········· 33
  - 3.2.3 Classical Swine fever ········· 35
  - 3.2.4 Porcine reproductive and respiratory syndrome (PRRS) ········· 36
  - 3.2.5 Newcastle disease ········· 36
- 3.3 Prevention and control of major zoonosis ········· 37
  - 3.3.1 Brucellosis ········· 37
  - 3.3.2 Bovine tuberculosis ········· 39
  - 3.3.3 Schistosomiasis ········· 39
  - 3.3.4 Echinococcosis ········· 40
  - 3.3.5 Rabies ········· 40
- 3.4 Eradication of glanders and Equine infectious anaemia ········· 41
- 3.5 Disease cleansing surveillance in breeder poultry farms and key original breeder pig farms ········· 42
- 3.6 Prevention of exotic animal diseases ········· 43
  - 3.6.1 Animal spongiform encephalopathies (BSE, scrapie) ········· 44
  - 3.6.2 African swine fever ········· 45
  - 3.6.3 Other exotic animal diseases ········· 46
- 3.7 Prevention & control of other terrestrial animal diseases ········· 47
  - 3.7.1 H7N9 influenza ········· 47

	3.7.2	Peste des petits ruminants ·· 48
	3.7.3	Porcine epizootic diarrhea ··· 49
	3.7.4	Other OIE listed diseases ·· 49

3.8 Aquatic animal disease prevention and control ·································· 51

- 3.8.1 Spring viraemia of carp (SVC) ·· 51
- 3.8.2 White spot syndrome (WSS) ·· 51
- 3.8.3 Infectious haematopoietic necrosis (IHN) ································· 52
- 3.8.4 Koi hepes-virus disease (KHVD) ··· 52
- 3.8.5 Cryptocaryon irritans ·· 52

3.9 Development of the mechanism of animal disease prevention & control ··· 53

- 3.9.1 Continuously improving emergency mechanism ······················· 53
- 3.9.2 Continuously improving designated liaison system for major animal disease prevention & control ································ 54
- 3.9.3 Continued zoning management ··· 55

Chapter 4 Veterinary Administrative Law Enforcement ················ 56

4.1 Animal Health Supervision ··· 56

- 4.1.1 Inspection of animals and animal products ······························ 56
- 4.1.2 Review of animal prevention epidemic conditions ··················· 57
- 4.1.3 Animal Health Supervision and Law Enforcement ·················· 58
- 4.1.4 Bio-safety disposal of animals dying of diseases ···················· 59
- 4.1.5 Clean-up and Rectification of Animal Medical Institutions ········ 61
- 4.1.6 Animal identification and animal disease traceability system ········ 61

4.2 Bio-safety Supervision of Veterinary Laboratories ·························· 63

Chapter 5　Veterinary Drug Production and Supervision 64

5.1　Production, Research & Development of Veterinary Drugs 64
5.2　Supervision of Veterinary Drug Production 65
 5.2.1　Veterinary Drug GMP Management 65
 5.2.2　Veterinary Drug Products QR Code Traceability Pilot 66
 5.2.3　Major Animal Disease Vaccine Quality Supervision 67
 5.2.4　Veterinary Drug Quality Supervision Sampling Inspection 67
5.3　Supervision in Veterinary Drug Operation 69
5.4　Supervision in Veterinary Drug Utilization 69
 5.4.1　Setting Up Prescription Drug Management System 69
 5.4.2　Strengthening Veterinary Drug Residue Monitoring 70
 5.4.3　Special Rectification on Veterinary Antimicrobial Drug 72
 5.4.4　Surveillance on Bacterial Drug Resistance 73
5.5　Veterinary Appliances Supervision 74

Chapter 6　Slaughtering Supervision 75

6.1　Boosting function adjustment of slaughtering supervision 75
6.2　Intensifying slaughtering supervision 76
6.3　Advancing transformation and upgrade of slaughtering industry 77
6.4　Promoting statistical monitoring, publicity and training of slaughtering industry 77

Chapter 7　Domestic and International Exchanges and Cooperation 79

7.1　Exchanges and Cooperation with International Organizations 79
 7.1.1　Deepen Exchanges and Cooperation with OIE 79
 7.1.2　Strengthen Exchanges and Cooperation with FAO 84

7.1.3 Exchange and Cooperation with the World Bank ················· 88

7.2 Bilateral and Multilateral Exchanges and Cooperation ············· 89

7.2.1 Exchanges and Cooperation with Neighboring Countries ············· 89

7.2.2 Exchanges and Cooperation between China and European Union ········ 91

7.2.3 Exchanges and Cooperation with Other Countries ············· 92

7.3 Exchanges and Cooperation with Hong Kong, Macao and Taiwan regions of China ················· 92

7.3.1 Hong Kong and Macao ················· 92

7.3.2 Taiwan ················· 93

Annex 1 National Veterinary Laboratory Status ············· 94

1. National Veterinary Reference Laboratory ················· 94
2. National Veterinary Diagnosis Laboratory ················· 95
3. National Veterinary Drug Residue Benchmark Laboratory ············· 96
4. National Veterinary Key Laboratory ················· 96
5. MOA Veterinary Key Laboratory ················· 97
6. OIE Reference Center ················· 101

Annex 2 Colleges and Universities with Veterinary Major ············· 103

Chapter 1
Veterinary System

The Ministry of Agriculture (MOA) is responsible for the administration of national veterinary system, and the national entry-exit inspection authority is in charge of entry-exit inspection of animal and animal products. To guarantee production safety of the breeding industry, animal production food safety, public health security and environmental safety, the Chinese government has constantly been improving veterinary administration and the building-up of relevant systems and mechanisms, notably by accelerating the implementation of the system of official and licensed veterinarians.

1.1　Veterinary agencies and organizations

1.1.1　Veterinary administrative agencies

MOA has a national Chief Veterinary Officer.

MOA has a Veterinary Bureau to be responsible for animal diseases prevention and control, animal epidemics management, animal health supervision and law enforcement, veterinarian administration and veterinary drug administration and inspection, administration of livestock and poultry slaughtering industry, and the administration of Chinese veterinarians. The Veterinary Bureau is composed of Division of General Affairs, Division of

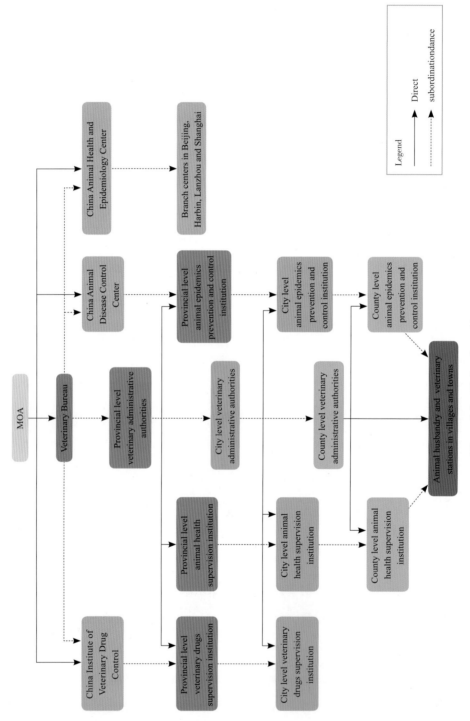

Figure 1-1　National veterinary system

Veterianrian Administration (Licensed Veterinarians Administration Office of MOA), Division of Science, Technology and International Cooperation, Division of Animal Epidemic Prevention, Division of Law Enforcement and Supervision, Division of Veterinary Drugs and Appliances Administration,and Division of Slaughtering Industry Administration. Please refer to http://www.syj.moa.gov.cn/jieshao/zhineng/ for detailed responsibilities of each division.

All provinces (autonomous regions and municipalities), cities and counties have their own veterinary administration authorities that are responsible for animal epidemic prevention, inspection, slaughtering administration, veterinarian administration and veterinary drug administration. By the end of 2014, veterinary administration authorities at province, city and county levels had a total of approximately 34,000 staff members.

1.1.2 Veterinary administrative law enforcement agencies

Local governments at county level or above have set up animal health supervision authorities, which are responsible for animal and animal products inspection as well as other supervisory and law enforcement work related to animal epidemic prevention. By the end of 2014, China had 32 provincial-level, 358 city-level and 3,162 county-level animal health supervision authorities and 22,681 branches of county-level agencies, employing nearly 150,000 staff members, in which 143,000 are responsible for law enforcement.

1.1.3 Veterinary technical support system

Veterinary technical support institutions provide technical support for animal diseases diagnosis and surveillance, epidemiological investigation, animal epidemic report and quality evaluation of veterinary drugs and biological products.

1.1.3.1 National veterinary technical support institutions

National veterinary technical support institutions include three public institutions directly under MOA, namely China Animal Disease Control Center, China Institute of Veterinary Drug Control and China Animal Health and Epidemiology Center.

China Animal Disease Control Center (Slaughtering Techniques Center of MOA) undertakes the country's animal diseases analysis and treatment, major animal diseases prevention and control, animal identification management, quality safety test of livestock and poultry products, as well as giving instructions on animal health supervision and providing technical support for livestock and poultry slaughtering supervision.

China Institute of Veterinary Drug Control (Veterinary Drug Evaluation Center of MOA) is responsible for veterinary drug evaluation, quality supervision and inspection of veterinary drugs and appliances, veterinary drug residue monitoring, strain (virus seed and insect egg) preservation, development and revision of national standards on veterinary drugs, and preparation and standardization of standard products and comparison products, etc.

China Animal Health and Epidemiology Center is responsible for epidemiological investigation, diagnosis and surveillance of major animal diseases, veterinary health evaluation, quality supervision and inspection of animal and animal products, animal health laws and standards, and prevention and control technologies research against exotic animal diseases. The Center, with four branch centers in Beijing, Harbin, Lanzhou and Shanghai, works with various levels and types of veterinary laboratories in conducting epidemiological investigations.

Besides, MOA has established 304 national animal epidemics surveillance and reporting stations and 146 border animal disease monitoring stations in border areas to conduct diseases monitoring and epidemiological investigations.

1.1.3.2 Local veterinary technical-support institutions

All provinces and the majority of cities and counties in China have established animal diseases prevention & control institutions to handle technical work regarding animal epidemics prevention and control such as surveillance, diagnosis and epidemiological investigations. By the end of 2014, such institutions employ a total of 35,000 staff members.

All provinces and some of the cities and counties in China have their own veterinary drug test institutions to test and supervise veterinary drugs. So far the country has 32 provincial level, 171 city level and 289 county level veterinary drug test institutions.

Veterinary departments within county-level governments have set up animal husbandry and veterinary stations in villages and towns and established teams for epidemics prevention in villages by buying their services to provide assistance in animal epidemics prevention and inspection and help organize non-profit technology promotions. Currently the country has 30,380 such veterinary stations with 645,000 employees.

1.1.4 Animal production management and technology promotion institutions

MOA has a Department of Livestock Production (National Feed Office) to be responsible for the administration of national animal husbandry and feed industries. Provinces, cities and counties have their own animal husbandry administrative authorities to be responsible for the administration of animal husbandry and feed industries within their jurisdiction and participate in national prevention and control of BSE and scrapie, etc.

The National Animal Husbandry Station under MOA is responsible for fine breed and technology promotion in the animal husbandry (including feed, prataculture and dairy industry), livestock and poultry varieties resources reservation and utilization management, quality control and certification in the animal husbandry, market discipline and industrial evaluation of the feed industry. Animal husbandry stations have been established in provinces, cities and counties.

1.1.5 Entry-exit inspection and quarantine system

The General Administration of Quality Supervision, Inspection and Quarantine (AQSIQ) is responsible for entry-exit inspection and quarantine of animal and animal products. The Department of Supervision on Animal and Plant Quarantine and Bureau of Import and Export Food Safety of the AQSIQ are responsible for the inspection and quarantine, supervision and management, and risk analysis of exiting and entering animals, plants, and related products, as well as import and export food and cosmetics respectively. The AQSIQ adopts vertical management over entry-exit inspection and quarantine institutions. It has established 35 entry-exit inspection and quarantine bureaus in provinces (autonomous regions and municipalities) and main ports, and nearly 300 branch bureaus and over 200 offices in land, sea and air ports and cargo distribution centers, totally employing over 30,000 persons.

1.1.6 Veterinary scientific research system

China's veterinary scientific research system operates on the central and local levels. There are nine central-level research institutions either under MOA or AQSIQ (table1-1). At the local level, most provinces have their own animal husbandry and veterinary research institutes to conduct such technical studies as on local prevention & control of animal diseases[①].

Table 1-1　Veterinary research institutes at the central level

Name	Main functions	Official website
Harbin Veterinary Research Institute, CAAS	Conduct researches in the prevention technologies and their basic theories of infectious animal diseases. Achievements have been made in etiology, epidemiology, pathogenesis, pathology, diagnostics, immunology and molecular biology in terms of infectious animal diseases. Especially well known for researches in animal flu, swine diseases and horse diseases	http://www.hvri.ac.cn
Lanzhou Veterinary Research Institute, CAAS	Specializes in the researches of serious fulminating infectious animal diseases and parasitic diseases. Conduct basic, applied and development researches in the pathogens, diagnosis, immunity and prevention of animal epidemics. Address issues about the sustainable control of major livestock and poultry diseases and guarantee healthy development of animal husbandry, long well known in the researches of herbivorous animal diseases (esp. foot-and-mouth disease)	http://www.chvst.com
Shanghai Veterinary Research Institute, CAAS	Conduct researches in forward-looking prevention technologies and relevant theories on livestock and poultry diseases and zoonosis which pose serious threats to animal husbandry	http://www.shvri.ac.cn
Institute of Animal Sciences of CAAS	Conduct basic applied researches and application and development researches in animal genetic resources and breeding, animal biotechnology and reproduction, animal nutrition and feed, pratacultural science and animal medicine. Address fundamental and key technological issues of national importance	http://www.iascaas.net.cn
Lanzhou Institute of Husbandry and Pharmaceutical Sciences of CAAS	Specialize in basic application researches and applied researches on veterinary drugs innovation, breeding and resources protection of herbivorous animals, modernization of Chinese veterinary medicine, and xerophyte pasture variety breeding and application	http://www.lzmy.org.cn

---

① *Veterinary Science and Technology Development in China (2013-2014)* for detailed information on veterinary research system and achievements.

(continued)

Name	Main functions	Official website
Institute of Special Animal and Plant Sciences of CAAS	Conduct researches on special economic animals and plants. Specialize in development, application and protection of rare, valuable and economic wild animal and plant resources. Conduct basic researches on application and development, especially on the applied technologies for domestic animals and plants	http://www.caastcs.com
Changchun Veterinary Research Institute, CAAS	Well known for researches in animal virology (esp. rabies), bacteriology, parasitology, animal production food safety, bio-toxicology, veterinary pharma-toxicology, and bio-safety technologies and equipment	http://cvrirabies.bmi.ac.cn
Chinese Academy of Inspection and Quarantine	Specialize in applied researches on inspection and quarantine. Also conduct fundamental, advanced technologies and soft science researches. Address comprehensive, key, unexpected and fundamental technological issues. Provide technical support for national decision-making on inspection and quarantine and offer technical assistance on law enforcement as requested by AQSIQ	http://www.caiq.org.cn
Yellow Sea Fisheries Research Institute, Chinese Academy of Fishery Sciences	Specialize in researches on the sustainable development and application of marine biological resources, including marine stock enhancement and aquaculture, fishery resources, and environmental and fishery engineering technologies	www.ysfri.ac.cn

To improve the ability of animal epidemics prevention & control and enhance technical support for the regulation on animal-derived food safety, MOA has set up 3 national veterinary reference laboratories, 4 national veterinary diagnostic laboratories, and 4 national veterinary drug residue benchmark laboratories. The Ministry of Science and Technology (MOST) has approved the establishment of 3 national key veterinary laboratories; MOA has designated 2 comprehensive key veterinary laboratories, 15 specialized/regional key veterinary laboratories and 12 observation stations on agricultural science. 15 reference centers have been recognized by OIE (Attachment 1).

1.1.7 Veterinary higher education system

There are 69 Chinese universities or colleges equipped with veterinary or animal medicine departments (Attachment 2), including China Agricultural University (CAU), producing nearly 7,000 graduates each year. Four among them are listed in the "985 program", namely, CAU(China Agricultural University), Northwest Agriculture & Forestry University, Shanghai Jiaotong University (SJTU) and Zhejiang University. They are also part of "211 program", together with 10 other universities, including Nanjing Agricultural University (NJAU), Huazhong Agricultural University (HZAU), Northeast Agricultural University (NEAU), Guangxi University (GXU), Tibet University, Southwest University, Sichuan Agricultural University, Qinghai University, Shihezi University and Ningxia University. Their veterinary or animal medicine departments not only produce veterinary professionals, but also constitute a significant part in China's veterinary scientific research system, playing an essential role in improving the country's veterinary technologies.

1.1.8 Societies, associations and professional technical committees

China has established NGOs(Non-government organizations) such as veterinary societies, associations and technical committees at both central and local levels so as to mobilize all powers at hand to improve animal health. The followings are at the national level:

Chinese Association of Animal Science and Veterinary Medicine was established in 1936 as a national academic group consisting of people working in animal science and veterinary medicine fields, engaged in domestic and international academic exchanges and promotes scientific and technological cooperation. It also provides technical consultation and services to the decision-making of national veterinary technical development strategies and relevant policy and economic buildings.

Chinese Veterinary Medical Association was established in October 2009, coordinating internal and external relationships of the industry, supporting law-based veterinary practice, enhancing self-discipline, improving professional ethics building, and standardizing the behaviors of licensed veterinarians.

Chinese Veterinary Drug Association was established in 1991 to promote industrial self-discipline, coordination, service and management. It is responsible for developing guidelines and rules for veterinary drugs industry and assisting the government in improving industrial management.

National Standardization Technical Committee of Animal Health was formerly known as National Animal Quarantine Standardization Technical Committee, and later National Animal Epidemic Prevention Standardization Technical Committee. It was first established in 1991, responsible for national technical standardization of animal health, covering such areas as animal epidemic prevention & control, animal products hygiene, animal health supervision, clinical diagnosis and treatment of animal diseases (excl. companion animals), animal welfare and performance evaluation of veterinary services.

National Standardization Technical Committee for Slaughtering and Processing was established in May 2011, responsible for standardardization in veterinary food hygienic quality and inspection, slaughter plants (houses) building, slaughter plants (houses) grading, designing of slaughter workshops and assembly lines, slaughtering and processing technologies, slaughtering and processing procedures and techniques, facilities and equipment for slaughtering and meat products processing equipment and techniques for disposal, and processing of inedible animal products, etc.

China Veterinary Pharmacopoeia Committee was established in 1986 as the statutory technical institution responsible for organizing the development and revision of national standards for veterinary drugs. It has six professional committees. The Committee is responsible for developing and revising national standards for veterinary drugs, compiling and publishing China Veterinary Pharmacopoeia, studying important topics related to veterinary drugs standards, reviewing drafts of national standards for veterinary drugs and relevant technical manuals, and participating in international technical exchanges held by professional organizations.

National Expert Committee on Veterinary Drug Residue was established in 1999 as a technical consultative institution for the monitoring of veterinary drug residue in animals and animal products. It is responsible for developing, reviewing and revising national monitoring plans on veterinary drug residue in animals and animal products, and evaluating the effect of monitoring plans; reviewing development and revision plans of national standards for veterinary drug residue and doing key researches on relevant topics; reviewing draft national standards of veterinary drug residue as well as relevant technical manuals for drug residue monitoring; participating in international technical exchanges held by relevant professional organizations; undertaking other tasks designated by the sub-committee on veterinary drug residue under the Food Safety National Standards Review Committee.

National Expert Committee on Animal Epidemic Prevention was established in November 2009 as an expert organization providing decision-making consultation and technical support for national animal epidemic prevention and control. It is responsible for discussing, studying and judging domestic and international epidemic situation and providing policy suggestions on prevention and control; evaluating existing prevention and control measures and offering advice on their adjustment and improvement; and providing consultation for important decision-making on prevention and control.

National Expert Committee on Animal Health Risks Assessment was established in November 2007 as an expert organization to evaluate animal health risks and provide consultation and technical support to the management of such risks. It is responsible for risk evaluation in major, exotic and emerging animal diseases, as well as risk assessment of animal health conditions and animal and animal products health and safety.

Veterinary Drug Review Committee of MOA was established in 1991 and has been managed as a review expert database since 2005, responsible for reviewing new veterinary drugs, new biological products and veterinary drug registration application from domestic and foreign enterprises, and re-evaluating approved veterinary drugs.

Veterinary Drug GMP Working Committee of MOA was established in 2001 as a statutory technical institution responsible for organizing the development and revision of GMP technical standards on veterinary drugs. Its duties include reviewing GMP specifications on veterinary drugs, GMP standards for veterinary drugs tests, examination and evaluation, and GMP guidelines on tests and examination of veterinary drugs; and reviewing disputes and appeals in GMP tests and examinations on veterinary drugs and offering solutions.

National Committee on Bio-safety of Pathogenic Microbe Labs. According to the *Regulation on Biosafety Management of Pathogenic Microbe Labs*, competent authorities under the State Council, together with experts from relevant State Council departments in etiology, immunology, laboratory medicine, epidemiology, preventive veterinary medicine, environmental protection and lab management, established the National Committee on Bio-safety of Pathogenic Microbe Labs. It is responsible for bio-safety evaluation and technical consultation and demonstration of the establishment and operation of labs running experiments related to highly pathogenic microbes.

1.2 Capacity building of veterinarians

1.2.1 Official veterinary

MOA has continued on qualification verification of official veterinarians and adjusts and replenishes the official veterinary team in time. It also keeps on offering training for official veterinarians, (such as the national enrichment class for official veterinarians). Under the instruction of MOA, local veterinary authorities are in charge of appointment and training of official veterinarians. By the end of 2014, China had verificated a total of more than 100,000 official veterinarians, trained 8,050 official veterinary teachers, and had trained a total of 104,000 official veterinarians. Besides, MOA has organized the compilation of *National Plan for Veterinary Team Building*, provinces, cities and counties have been offering training programs accordingly to enhance the quality of teams.

1.2.2 licensed veterinarians

National licensed veterinarian qualification examination. MOA is responsible for the organization of national licensed veterinarian qualification examination. It is constantly improving the management of the examination and trying new exmination methods, for example by shifting from pre-examination review to post-examination review. Pilot examination were first held in Tibet Autonomous Region for Type-C licensed veterinarians, and the national examination committee was responsible for determining the specific pass/fall line. In 2014, a total of 16,143 were accredited as licensed veterinarians in China, in which 8,280 were veterinarians and 8,063 were assistant veterinarians.

Licensed veterinary qualification conferring. According to the *Regulations on Practicing Veterinarians* and *Notice of MOA on Qualification Review and Accreditation of Practicing Veterinarians* (Nongyifa [2013] No.15), MOA is responsible for qualification review, granting, examination and approval of licensed veterinarians. In 2014, the MOA granted 6,571 persons with senior titles as qualified licensed veterinarians.

1.2.3 Rural veterinarian

China adopts registration system for rural veterinarians. By the end of 2014, registered rural veterinarians reached 277,000.

Village epidemic prevention workers enjoy subsidies from both central and local governments. In 2014, the central government allocated RMB780 million for grassroots animal epidemics prevention. Provinces have been increasing the amount of subsidy on an annual basis based on their financial conditions and resources.

1.3 PVS evaluation

Performance of Veterinary Services (PVS) evaluation was established by OIE as a comprehensive international standard to evaluate the veterinary management system and its performance and quality level of a country (region). It emphasizes the objective evaluation of the overall capacity of veterinary system to constantly improve the overall performance of the system.

In 2014, MOA required all provinces to conduct PVS self-evaluation based on OIE tool. Four parts were included: 1) manpower, material resources and financial resources; 2) technical capacity; 3) shareholder interaction; and 4) market access, and they were detailed into 46 key indicators. According to the results, China's building of veterinary institutions, the capability of persons in charge of competent veterinary authorities, veterinary institutions stability, and policy consistency were satisfactory.

Chapter 2
Veterinary Laws and Regulations

The Chinese government values the building of veterinary laws and regulations system. In recent years, the system has been constantly improved. The country has established a legal framework centering around *Law of the P. R. China on Animal Epidemic Prevention* and the *Law of P. R. China on Entry & Exit Animal and Plant Quarantine,* and complemented by administrative regulations, departmental rules, and local laws and regulations which are necessary to improving veterinary services. Currently there are four laws, eight administrative laws and regulations, over 30 support rules and technical manuals (Table 2-1).

Table 2-1 Veterinary laws and regulations system

Category		Name	Date of implementation	Main contents
Laws and regulations	Laws	Law of the P. R. China on Animal Epidemic Prevention	Jan 1st, 2008	Rules about the prevention of animal epidemics, reporting and announcement of animal epidemic situations, control and extinguishing of animal epidemics, inspection of animal and animal products, animal diagnosis and treatment, and supervision and management of animal epidemics prevention
		Law of the P. R. China on Entry &Exit Animal and Plant Quarantine	April 1st, 1992	Specific rules about quarantine review and approval, entry, exit, transit and means of transport, as well as quarantine measures on carried and mailed things

Chapter 2 Veterinary Laws and Regulations

(continued)

Category		Name	Date of implementation	Main contents
Laws and regulations	Laws	Law of the P. R. China on Animal Husbandry	July 1st, 2006	Specific rules about the protection of animal genetic resources; variety breeding and production of breeder animals; livestock and poultry farming, trading and transport; and animal products quality safety guarantee
		Law of the P.R. China on Quality and Safety of Agricultural Products	Nov 1st, 2006	Specific rules about quality safety standards, producing area, production, packaging and labeling, supervision and inspection of agricultural products
	Regulations of the State Council	Regulation on Emergency Response to Major Animal Epidemic	Nov 18th, 2005	Detailed rules about the emergency disposal principles, emergency preparedness, surveillance, reporting and publishing, emergency disposal and legal liabilities of major animal epidemics
		Regulation on Bio-safety Management of Pathogenic Microbe Labs	Nov 5th, 2004	Specific rules about classification and management of pathogenic microbes, establishment and management of laboratories, and lab infection control and supervision
		Regulations on Administration of Veterinary Drugs	Nov 1st, 2004	Specific rules about the development, production and approval of new veterinary drugs; production, operation, use, and import and export of veterinary drugs; labeling and advertising of veterinary drugs; supervision and administration of veterinary drugs
		Regulation on Supervision and Administration of the Quality and Safety of Dairy Products	Oct 6th, 2008	Specific rules about farming, diseases prevention and treatment, production and purchasing, and dairy stock transport
		Regulations on Administration of Slaughtering of Pigs	Aug 1st, 2008	Rules about appointed-location slaughter and centralized quarantine of pigs and their supervision and administration
		Regulations for the Implementation of the Law of P. R. China on Entry & Exit Animal and Plant Quarantine	Jan 1st, 1997	Specific rules about quarantine approval, entry quarantine, exit quarantine, transit quarantine, quarantine of carried and mailed contents, and quarantine of transport facilities

(continued)

Category		Name	Date of implementation	Main contents
Laws and regulations	Regulations of the State Council	Regulation on the Administration of Feeds and Feed Additives	May 1st, 2012	Rules about examination, approval, registration, production, operation and application of feeds and feed additives
		Regulations for the Administration of Affairs Concerning Experimental Animals	Nov 14th, 1988	Rules about the rearing of experimental animals, quarantine and infectious diseases control of experimental animals, application, import and export of experimental animals, and employees management
Departmental rules	Epidemics surveillance and reporting	Administrative Measures for Animal Epidemics Reporting	Oct 19th, 1999	Specific rules on the responsible parties and system, methods, contents and timing of animal epidemics reporting
		List of Category Ⅰ, Ⅱ and Ⅲ Animal Diseases	Dec 11th, 2008	List of 17 category Ⅰ animal diseases, 77 category Ⅱ and 63 category Ⅲ
		List of zoonoses	Jan 19th, 2009	List 26 diseases as notifiable zoonoses, including BSE, HPAI, rabies, anthrax and brucellosis, etc
		Regulatory Rules of National Surveillance and Reporting System for Animal Epidemics (trial)	June 10th, 2002	Detailed rules about the organization, responsibilities, targets, diseases and methods of surveillance, reporting of results, and management of various levels of surveillance and reporting centers, etc
	Emergency management	National Contingency Plan for Major Animal Epidemics	Feb 27th, 2006	Rules about grading of unexpected major animal epidemics, system and responsibilities of emergency organization, monitoring, early-warning and reporting of unexpected major animal epidemics, emergency response and termination, post-event measures and emergency disposal safeguard, etc
		National Contingency Plan for Highly Pathogenic Avian Influenza	Feb 3rd, 2004	Rules about the reporting, confirmation and grading, contingency system, control measures and safeguard measures of HPAI

(continued)

Category		Name	Date of implementation	Main contents
Departmental rules	Emergency management	Contingency Plan for Entry-Exit Major Animal Epidemics	June 30th, 2005	Rules about the emergency disposal of cases or suspected cases of major animal epidemics occurring or prevailing in China or abroad as well as those detected or discovered during entry-exit inspection and quarantine
		Contingency Plan for Prevention and Control of Foot-and-mouth Disease	March 27th, 2010	Rules about the emergency management measures for foot-and-mouth diseases, including prevention and emergency preparedness, monitoring and early-warning, emergency response and post-event rebuilding, etc
		Contingency Plan for Prevention and Control of Peste Des Petits Ruminants	June 6th, 2010	Rules about the surveillance, reporting, confirmation and grading of peste des petits ruminants as well as response, disposal and safeguard measures
		Contingency Plan for Prevention & control of Equine Influenza	April 15th, 2008	Rules about the surveillance, reporting, confirmation and grading of equine influenza as well as response, disposal and safeguard measures
		Contingency Plan of Agricultural Authorities for Human Infection of HPAI	Nov 15th, 2005	Rules about the early-warning and surveillance, and emergency disposal and response of outbreaks of human infection of HPAI
	Bio-safety management of veterinary labs	List of Animal Pathogenic Microbes Categories	May 24th, 2005	Four categories of animal pathogenic microbes, including ten category I microbes and eight category II
		Administrative Measures on the Preservation of Animal Pathogenic Microbe Strains (Virus Seeds)	Jan 1st, 2009	Rules about collection, preservation and supply, destroy, external exchanges and preservation institutions of animal pathogenic microbe strains (virus seeds)

(continued)

Category		Name	Date of implementation	Main contents
Departmental rules	Bio-safety management of veterinary labs	Management and Approval Measures for Bio-safety of Highly Pathogenic Animal Pathogenic Microbe Labs	May 20th, 2005	Rules about qualification approval, experimental activities and transport approval of highly pathogenic microbes labs
	Inspection and supervision administration	Administrative Measures for Animal Inspection	Mar 1st, 2010	Rules about the inspection application, inspection of producing area, slaughter inspection and supervision of animal and animal products (including aquatic offspring seeds and breeder dairy animals)
		Review Measures for Animal Epidemic Prevention Conditions	May 1st, 2010	Rules about epidemic prevention conditions, approval and licensing, supervision and management of animal farms, breeding villages, animal quarantine sites, animal slaughter and processing plants, disposal sites for animal and animal products, and markets
		Administrative Measures on Animal Identification and Farming Records	July 1st, 2006	Rules about the management of animal identification, farming records, relevant information management and supervision
		Administrative Measures on Roadside Animal Epidemic Prevention Supervision Checkpoints	Aug 28th, 2006	Rules about the set-up and supervision of roadside animal epidemic prevention checkpoints across the country
		Administrative Measures on Reporting of Animal Health Supervision Information (trial)	Jan 8th, 2007	Rules about the collection and reporting of information regarding animal health supervision and law enforcement
	Zoning management	Assessment and Management Measures for Specified Animal Diseases Free Zones	Mar 1st, 2007	Rules about application, assessment and announcement of zones free from specified animal diseases

(continued)

Category		Name	Date of implementation	Main contents
Departmental rules	Zoning management	Technical Norms for the Management of Specified Animal Diseases Free Zones	Jan 25th, 2007	Requirements on the establishment, standards, foundation and system, prevention and surveillance, inspection and supervision, emergency response and recovery of zones free from specified animal diseases
		On-site Review Form for Specified Animal Diseases Free Zones	Dec 9th, 2008	96 assessment items for specified animal diseases free zones
		Technical Norms for Surveillance of HPAI for Specified Animal Diseases Free Zones	Dec 1st, 2011	Rules about the basic requirements, method, and result interpretation of HPAI monitoring in specified animal diseases free zones, as well as the surveillance requirements for the proving and recovery of HPAI-free status
		Technical Norms for Surveillance of Foot-and-mouth Disease for Specified Animal Diseases Free Zones	Dec 1st, 2011	Rules about the basic requirements, method, and result interpretation of foot-and-mouth disease surveillance in specified animal diseases free zones, as well as the surveillance requirements for the proving and recovery of FMD-free status
		16 Technical Norms for Specified Animal Diseases Free Zones, Including Equine Influenza	Feb 23rd, 2009	Requirements on specified animal diseases free zones of 16 diseases, including equine influenza
		General Norms on the Building of Specified Animal Diseases from Compartments for Meat Poultry(Trial)	June 22nd, 2009	Rules about the general requirements, bio-safety management system, and supervision on veterinary authority for poultry compartment for meat poultry
		Standards for Specified Animal Diseases Free Compartments for Meat Poultry(Trial)	June 22nd, 2009	Standards for avian influenza free compartments for meat poultry, and rules about the diseases-free status of compartments and the revocation and restoration of the status

(continued)

Category		Name	Date of implementation	Main contents
Departmental rules	Zoning management	On-site Review Form for Specified Animal Diseases Free Compartments for Meat Poultry	June 7th, 2010	87 review items of compartments
	Veterinary drugs administration	Administrative Measures on the Operation of Veterinary Biological Products	May 1st, 2007	Rules about the distribution, operation and supervision of veterinary biological products
		Measures on Veterinary Drugs Registration	Jan 1st, 2005	Rules about the registration of new veterinary drugs, imported veterinary drugs, and changes in veterinary drugs, re-registration of imported veterinary drugs, reexamination, inspection and standard substance management of veterinary drugs
		Administrative Measures on the Approval Documents Number of Veterinary Drugs	Jan 1st, 2005	Rules about the application, approval, issuance and supervision of the approval documents number of veterinary drugs
		Administrative Norms on Production Quality of Veterinary Drugs	June 19th, 2002	Rules about the basic principles of production and quality management of veterinary drugs, the entire process of veterinary drug preparations production, and key production processes of crude drugs which could affect the quality of final products
		Administrative Measures on Veterinary Drugs Import	Jan 1st, 2008	Rules about the application of veterinary drugs import, and the operation and supervision of imported veterinary drugs
		Administrative Measures on New Veterinary Drugs Development	Nov 1st, 2005	Rules about the pre-clinical study, clinical trial and supervision of new veterinary drugs

(continued)

Category		Name	Date of implementation	Main contents
Departmental rules	Veterinary drugs administration	Acceptance Check Procedure for the Implementation of Administrative Measures on Production Quality of Veterinary Drugs	Sep 1st, 2010	Rules about the application and review, on-site inspection, approval and certification, and inspector management of veterinary drugs production
		Administrative Norms on Operation Quality of Veterinary Drugs	Mar 1st, 2010	Rules about the site and equipment, institution and personnel, rules and regulations, purchasing and warehouse delivery, display and storage, sales and transport, and after-sales service of veterinary drugs operation
		Administrative Measures on Labels and Instructions of Veterinary Drugs	July 1st, 2004	Rules about the basic requirements and management of labels and instructions of veterinary drugs
		Sampling Rules on Supervision and of Veterinary Drugs Quality	Dec 10th, 2001	Rules about the institution, personnel, sampling size, principles, requirements and precautions of quality supervision and sampling of veterinary drugs
		Administrative Measures on Prescription and OTC Veterinary Drugs	Mar 1st, 2014	Veterinary drugs are categorized into Prescription and OTC based on safety and risk considerations and are managed accordingly. And rules are provided for their management
		Directory of Basic Medicine Used by Rural Veterinarians	Mar 1st, 2014	Rules about the prescription veterinary drugs available to rural veterinarians engaged in animal diagnosis and treatment
	Management of feeds and feed additives	Administrative Measures on the Safety and hygieneof Animal-derived Feed Products	July 1st, 2012	Rules about the approval of establishment, production management, operation, import and usage management, and check and supervision of animal-derived feed enterprises
		Administrative Measures on Registration of Imported Feeds and Feed Additives	July 1st, 2014	Rules about registration review, test and supervision of imported feeds and feed additives

(continued)

Category		Name	Date of implementation	Main contents
Departmental rules	Management of feeds and feed additives	Administrative Measures on Supervision, Inspection and Quarantine of Import& Export Feeds and Feed Additives	Sep 1st, 2009	Rules about risk management, registration, inspection and quarantine, and supervision of import & export feeds and feed additives
	Veterinarian administration	Administrative Measures on Licensed Veterinarians	Jan 1st, 2009	Rules about qualification examination, registration, records keeping and activities management of licensed veterinarians
		Administrative Measures on Rural Veterinarians	Jan 1st, 2009	Rules about the registration and engagement in animal diagnosis and treatment of rural veterinarians
		Administrative Measures on Animal clinics	Jan 1st, 2009	Rules about the requirements (permission for diagnosis and treatment) and activities management of animal clinics
		Interim Procedures of the Qualification Test for licensed Veterinarians	Mar 1st, 2009	Rules about the organization, personnel, registration procedure, test content, results announcement, and violations management of the qualification test for licensed veterinarians
	Inspection procedures and technical norms	Inspection Procedures for Pigs/Ruminants/ Poultry/ Equine at Producing Areas	April 20th, 2010	Rules about the inspection targets, eligibility criteria, procedure, results handling and record-keeping for pigs, ruminants, poultry and equine animals at producing areas
		Inspection Procedures for Canine/Feline/ Rabbits at Producing Areas	Oct 19th, 2011	Rules about the inspection targets, eligibility criteria, procedure, results handling, record-keeping and protection for producing areas of dogs, cats and rabbits
		Inspection Procedures for Fish/Crustacean/Shellfish at Producing Areas (trial)	Mar 17th, 2011	Rules about the inspection targets, eligibility criteria, procedure, results handling and record-keeping for producing areas of fish, crustacea and shellfish

(continued)

Category		Name	Date of implementation	Main contents
Departmental rules	Inspection procedures and technical norms	Slaughtering Inspection Procedures of Pigs/Poultry/Cattle/Sheep/Goats	May 31st, 2010	Rules about the supervision and examination, inspection application, ante-mortem, simultaneous inspection, inspection results handling and record-keeping of pigs, poultry, cattle and sheep/goats in slaughterhouses (plants, sites)
		Bees Inspection Procedures	Oct 13th, 2010	Rules about the targets, eligibility criteria, procedures, results management and record-keeping of bees inspection
		Technical Norms for Prevention and Control of 14 Animal Diseases, including HPAI	April 9th, 2007	Rules about the prevention and control techniques of 14 animal diseases, namely, HPAI, FMD, EIA, glanders, brucellosis, TB, pseudo-rabies, CSF, ND, IBD, MD, sheep pox, anthrax and ALV-J, including operation procedure, technical standards and protection measures in confirmation, response, surveillance, vaccination, inspection and supervision
		Technical Norms for Prevention and Treatment of Highly Pathogenic Porcine Reproductive and Respiratory Syndrome	Mar 28th, 2007	Rules about the operation procedure and technical standards of the diagnosis, reporting, handling, prevention control, inspection and supervision of highly pathogenic porcine reproductive and respiratory syndrome
		Technical Norms for the Prevention and Control of Rabies	Oct 30th, 2006	Rules about the diagnosis, surveillance reporting, handling, prevention and control of animal rabies
	Entry-exit administration of animal and animal products	List of Category Ⅰ and Ⅱ Animal Infectious Diseases and Parasitic Diseases for the imported from others countries	Nov 28th, 2013	Names of 15 category Ⅰ, 147 category Ⅱ and 44 other infectious diseases and parasitic diseases which are subject to entry quarantine

(continued)

Category		Name	Date of implementation	Main contents
Departmental rules	Entry-exit administration of animal and animal products	Administrative Provisions on Risk Analysis of Entry Animal and Animal Products	Feb 1st, 2003	Rules about the hazard factors identification, risk assessment, risk management and risk commiunication of entry animal and animal products
		Administrative Measures on Quarantine Sampling of Entry-exit Animal and Animal Products	June 27th, 1992	Rules about the sampling sizes, requirements and standards of entry-exit animal and animal products
		Administrative Measures on Inspection, Supervision and Quarantine of Import and Export Meat Products	Jan 4th, 2011	Rules about import, export and transit inspection and quarantine as well as supervision of meat products
		Administrative Measures on the Quarantine Approval of Entry (Transit) Animal and Animal Products	July 30th, 2008	Rules about the quarantine application and approval of entry and transit animals and animal products
		Administrative Measures on Inspection, Supervision and Quarantine of Import and Export Aquatic Products	Jan 4th, 2011	Rules about the entry and exit inspection and quarantine and supervision of aquatic products
		Administrative Measures on Temporary Quarantine Sites for Entry-exit Animals	Nov 27th, 1996	Rules about the conditions and permission of temporary quarantine sites of entry-exit animals
		Administrative Measures on Quarantine of Genetic Materials of Entry Animals	July 1st, 2003	Rules about the quarantine and supervision of the genetic materials (semen, embryo and oocytes of mammals) of entry animals

(continued)

Category	Name	Date of implementation	Main contents
Depart-mental rules	Entry-exit administration of animal and animal products	Administrative Measures on the Application of Quarantine Sites for Entry Animals	Dec 1st, 2009	Rules about the supervision of national quarantine sites for entry animals
		Administrative Measures on the Quarantine of Belongings of Entry-exit Persons	Jan 1st, 2004	Rules about the animals, plants, and animal and plant products carried by entry-exit persons
	Technical Norms for IT application in veterinary health	Code Standards for IT Application in Veterinary Health (trial), DS Model Standards for IT Application in Veterinary Health (trial), Data Dictionary Standards for IT Application in Veterinary Health (trial), and Data Exchange Format Standards for IT Application in Veterinary Health (trial)	Dec 12th, 2014	Rules about the code standards and data model of data-set of veterinary health IT system, categorized information word data, data model information, data element information and code information of the master data-set of veterinary health business information, and the data exchange format of veterinary health IT system among different levels, nodes and subsystems

In 2014, to ensure implementation of relevant laws and regulations and identify problems and their causes, MOA kept on doing post-legislation assessment of *Law of P. R. China on Animal Epidemic Prevention* after its amendment in 2007, and also initiated assessment of *Administrative Measures on licensed Veterinarians* and *Administrative Measures on Animal clinics*. The effort was intended to assess the enforcement of these laws and regulations in terms of effectiveness, coordination, appropriateness and operability, so as to analyze the problems and their causes and offer suggestions for improvement accordingly.

Chapter 3

Prevention and Control of Animal Disease

Thanks to continued efforts in implementing the *National Medium and Long Term Plan for Animal Epidemics Prevention & control (2012-2020)*, no major regional animal epidemics occurred in 2014, and the production safety of farming industry, safety of animal-derived food and public health security were assured. In 2014, animal husbandry produced RMB2.9 trillion in China, 28.3% of that of agriculture, forestry, animal husbandry and fishery. Meat production reached 87,067,400 tons, an increase of 2.01% from 2013. Eggs and milk production totaled 28,938,900 and 37,246,400 tons, a year-on-year increase of 0.62% and 5.47% respectively.

3.1 OIE listed diseases that were never reported or already eradicated in China

China eradicated rinderpest in 1955 and was recognized by OIE as a country free from rinderpest in 2008. Contagious bovine pleuropneumonia was eradicated in 1996, and China was recognized by the OIE free from the disease in 2011. Bovine spongiform encephalopathy (BSE) and African horse sickness have never occurred in China, and the country

Chapter 3 Prevention and Control of Animal Disease

was recognized by OIE as one with negligible BSE risk and free from African horse sickness in 2014. Table 3-1 showed OIE listed diseases that was never occurred in China.

Figure 3-1　OIE certificates granted to China

Table 3-1 OIE listed diseases which never occurred in China

Susceptible animal species	Diseases
Multiple species	Crimean-Congo hemorrhagic fever, heartwater, New world screwworm, Old world screwworm, Q fever, Rift valley fever, surra (Trypanosome evansi), Tularemia, Vesicular stomatitis, West Nile fever, Echinococcosis multilocularis
Cattle diseases	Bovine spongiform encephalopathy, lumpy skin disease
Sheep and Goat diseases	Maedi-visna disease, Nairobi disease, Scrapie
Swine diseases	African swine fever, Nipah virus Encephalitis
Horse diseases	African horse sickness, Dourine, Equine encephalomyelitis (eastern), Equine encephalomyelitis (western), Equine viral arteritis, Equine piroplasmosis, Venezuelan Equine Encephalormyelitis
Rabbit disease	Myxomatosis
Bee diseases	Acarapisosis of honey bees, American foulbrood of honey bees, European foulbrood of honey bees, Tropilaelaps infestation of honey bees, varroosis of honey bees
Fish diseases	Epizootic haematopoietic necrosis, Red sea bream iridovirus disease, Gyrodactyliasis, Viral haemorrhagic septicaemia, Infectious salmon anaemia, Epizootic ulcerative syndrome
Mollusc diseases	Infection with *Bonamia Ostreae*, Infection with *Perkinsus Olseni*, bonamia protozoon infection, abalone wilting syndrome, Infection with *Marteilia Refringens*, Infection with abalone herpes-like virus, Infection with *Perkinsus Marinus*
Shellfish diseases	Crayfish plague, Infectious myonecrosis, Necrotizing hepatopancreatitis
Amphibian diseases	Infection with Ranavirus, Infection with *Batrachochytrium Dendrobatidis*
Others	Camel pox, Leishmaniasis

3.2 Prevention and control of major animal diseases

3.2.1 Foot-and-mouth disease

China takes the following measures to prevent and control foot-and-mouth disease:

Compulsory vaccination. In 2014, China continued to implement compulsory vaccination against FMD and conduct monitoring of the vaccination effect. All pigs were compulsory vaccinated against type

O FMD; all cattle, sheep, camels and deer were compulsory vaccinated against type O and Asian type I FMD; all dairy cattle and breeding oxen were compulsory vaccinated against type A FMD; all cattle, goat and sheep in border areas in Guangxi, Yunnan, Tibet, Xinjiang and Xinjiang Production and Construction Corps were compulsory vaccinated against type A FMD. Free-ranging livestock are subject to centralized vaccination once in spring and again in fall. New stock entrants should be vaccinated in a timely manner. In 2014, China consumed 3.47 billion ml FMD vaccine, and the average immunization density stayed above 95%. The overall average qualification rate of immunity antibody reached 87.81%, and that of type O, Asian type I and type A was 87.48%, 88.25% and 89.88% respectively.

Surveillance and epidemiological investigation. Surveillance and epidemiological investigation of foot-and-mouth disease are carried out according to the requirements in the *National Plan for Animal Disease Surveillance and Epidemiological Investigation 2014* (Nongyifa〔2014〕No.2) and the *Foot-and-mouth Disease Surveillance Plan*. Various levels of animal diseases prevention and control institutions are responsible for surveillance of the breeder farms, large-scale farms, free-ranging farming households, livestock trading markets and slaughter houses of such cloven-hoofed animals as pigs, cattle, goat and sheep. National FMD reference labs enhanced etiological surveillance and molecular epidemiological comparative analysis, tracking virus variation, and especially monitoring livestock in high-risk areas which have had FMD outbreaks as well as in border areas. Surveillance in FMD free zone where vaccination is not practised has been enhanced. Veterinary authorities and national FMD reference labs conduct regular surveillance on the farming and distribution in such areas as well as the pigs, cattle, sheep, and other susceptible animals and their products entering these areas. In 2014, China tested a total of 4,710,000 FMD samples (3,829,000 serological samples and 881,000 etiological samples), in which 6 and 14 were positive for type O and type A FMD respectively. Immediate measures were taken to handle positive livestock and those in the same herds.

Zoning management. In July 2014, a forum was held in Heilongjiang on the building of disease-free zone in the great northeast. Given the special geographic advantages, rich natural resources and favorable conditions for animal epidemics prevention in the northeast, participants of the forum made a plan to adopt FMD zoning management in Heilongjiang, Jilin, Liaoning and eastern Inner Mongolia. Following the principle of "each province takes care of building simultaneously under national plan, accepts independent assessment, and links all four areas together as a whole", the target is to establish "great northeast" (three North-eastern provinces and eastern Inner Mongolia) FMD-free zone by the end of 2018. The forum also exploited such specific issues as the institutional framework and time schedule. In September 2014, the *Guidance and Implementation Plan for the Building of Great Northeast FMD-free Zone* (draft) was drafted at a liaison conference held in Qingdao.

Emergency disposal. In 2014, seven FMD outbreaks were reported in Tibet, Jiangsu and Jiangxi province, involving 74 cases (7 pigs and 67 cattle), and 324 were killed (44 pigs and 280 cattle). According to the requirements in relevant contingency plans and prevention and control technical standards, efforts were made to prevent and control the epidemics according to law and in a scientific manner. Epidemic areas were strictly quarantined. Disinfection for source elimination, monitoring and close examination were conducted. Sick animals and those in the same herds were killed or put under biosafety disposal. Thanks to these efforts, the epidemics were effectively controlled and eliminated without further spread (Table 3-2).

Table 3-2　FMD in China, 2014

Province	Outbreaks	Species	Serotype	Cases	Deaths	Slaughtered
Tibet	4	Cattle	Type A	61	0	262
Jiangsu	2	Pig	Type A, type O	7	0	44
Jiangxi	1	Cattle	Type O	6	0	18
Total	**7**	**Pig, cattle**	**Type A, type O**	**74**	**0**	**324**

3.2.2　HPAI

China takes the following measures to prevent and control HPAI:

Compulsory vaccination. In 2014, China kept on adopting compulsory vaccination against HPAI and monitoring the effect. A total of 15.8 billion doses of HPAI vaccines were used, the average immunity density stayed above 95%, and the overall average qualification rate of immunity antibody reached 91.9%。

Surveillance. On March 20, 2014, MOA released the *National Plan for Animal Diseases Surveillance and Epidemiological Investigation 2014* (Nongyifa〔2014〕No.2), including the *Surveillance Plan for Animal Influenza*. Immune antibody and etiological surveillance were conducted on chicken, duck, goose and other poultry, wild birds, minks, raccoon dogs and other economic animals, as well as artificially fed wild animals such as tigers in breeder poultry farms, commercial poultry farms, scattered farming households, live poultry trading markets, slaughter houses, main habitats of migratory birds and key border areas, and also on pigs and environmental samples in high-risk areas. In 2014, H5 sub-type avian influenza surveillance was conducted on poultry and pigs in 8,773 breeder poultry farms, 72,379 commercial farms, 49,483 scattered farming households, 8,246 trading markets, 264 wild bird habitats, 811 other sites and one slaughter house; 3,860,000 influenza samples from various animals were tested, including 3,340,000 serological samples. Among the 520,000 etiological samples tested, 159 were positive for H5 sub-type avian influenza, including 30 H5N1, 19 H5N2, 1 H5N3, 104 H5N6 and 5 H5N8 sub-types. Immediate measures were taken on positive poultry and those in the same herds according to relevant rules. Besides, national reference labs enhanced etiological monitoring and molecular epidemiological analysis on avian influenza and kept on tracking virus variation.

Emergency disposal. In 2014, three H5N1 outbreaks occurred in Hubei, Guizhou and Yunnan province, and one H5N6 outbreak occurred in Heilongjiang, involving 60,500 cases, 51,600 deaths and 5,280,000 destroyed (Table 3-3). According to relevant contingency plans and prevention and control technical standards, local governments handled these outbreaks in a law-based and scientific manner. Epidemic areas were strictly quarantined. Sick animals and those in the same herds were killed or put under biosafety disposal. Disinfection for source elimination, monitoring and close examination were enhanced. Thanks to these efforts, epidemics were effectively controlled and eliminated without further spread.

Table 3-3　HPAI in China, 2014

Location	Sub-type	Species	Cases	Deaths	Slaughtered
Yangxin County, Huangshi, Hubei province	H5N1	Laying hen	6,700	3,200	68,906
Xixiu District, Anshun, Guizhou province	H5N1	Laying hen	3,629	976	323,292
Tonghai County, Yuxi, Yunnan province	H5N1	Laying hen	29,600	29,600	4,823,085
Shuangcheng District, Harbin, Heilongjiang province	H5N6	Goose	20,550	17,790	68,884
Total			60,479	51,566	5,284,167

3.2.3　Classical Swine fever

All pigs are subject to compulsory vaccination with live swine fever vaccine and live passage cell-derived swine fever vaccine, and the effect was monitored. Local governments have strictly followed the state requirements of compulsory vaccination. In 2014, a total of 1,590 million doses of classical swine fever vaccines were used in China.

Immune antibody and etiological surveillance are conducted according to the requirements in the *Swine Fever Surveillance Plan* under the *National Plan for Animal Diseases Surveillance and Epidemiological Investigation 2014* (Nongyifa〔2014〕No.2), especially on pigs in breeder pig farms, medium and small-sized farms, trading markets, slaughter houses and areas which had experienced classical swine fever outbreaks. In 2014, 1,843,300 swine fever samples were tested in China, including 121,700 etiological samples, and 224 came out positive. Immediate measures were taken on positive pigs and those in the same herds.

In 2014, 28 classical swine fever outbreaks took place in five provinces (autonomous regions), namely Yunnan, Guizhou, Gansu, Shaanxi and Guangxi, involving 837 sick and 138 dead pigs (Table 3-4).

Table 3-4　Classical Swine fever in China, 2014

Province	Outbreaks	Cases	Deaths
Guangxi Zhuang Autonomous Region	12	643	92
Guizhou	1	4	0
Yunnan	8	169	37
Shaanxi	4	15	5
Gansu	3	6	4
Total	**28**	**837**	**138**

3.2.4 Porcine reproductive and respiratory syndrome (PRRS)

MOA continued guiding local governments to conduct vaccination and surveillance against PRRS. In 2014, a total of 2.39 billion ml PRRS vaccine were used in the country. Immune antibody and etiological surveillance were conducted particularly on pigs in breeder pig farms, medium and small-sized farms, trading markets, slaughter houses and areas which had experienced PRRS. In 2014, ten provinces (municipalities or autonomous regions) reported outbreaks of PRRS, namely Guangxi, Yunnan, Jiangxi, Hubei, Zhejiang, Anhui, Sichuan, Shaanxi, Chongqing and Hebei, involving 1,347 sick pigs and 162 were killed. Also, one highly pathogenic PRRS outbreak was reported in Guizhou.

3.2.5 Newcastle disease

MOA continued guiding local governments to conduct vaccination and surveillance of Newcastle disease according to the requirements in *Newcastle Disease Surveillance Plan*, especially on poultry (chicken, duck, goose, turkey, pigeon and quail) in breeder poultry farms, commercial poultry farms, and live poultry markets. Also, according to the requirements of the *National Plan for Animal Disease Surveillance and Epidemiological Investigation 2014*, the national Newcastle disease diagnosis labs collected 15,443 cotton swab poultry samples in 217 sites from 40 counties or districts in 13 provinces and municipalities, including Jiangsu and Anhui, and conducted molecular epidemiological analysis on 563 isolated Newcastle disease virus strains.

In 2014, 95 Newcastle disease outbreaks took place in eight provinces (autonomous regions), namely Guangxi, Jiangxi, Yunnan, Guizhou, Hubei, Gansu, Shaanxi and Zhejiang province, involving 14,000 cases, 6,459 deaths and 4,223 Slaughtered (Table 3-5).

Table 3-5 Newcastle disease in China, 2014

Province	Outbreaks	Cases	Deaths
Zhejiang	1	17	2
Jiangxi	67	4,427	4,129
Hubei	1	450	125
Guangxi Zhuang Autonomous Region	12	6,230	1,301
Guizhou	1	62	62
Yunnan	8	927	409
Shaanxi	1	1,500	25
Gansu	4	416	406
Total	**95**	**14,029**	**6,459**

3.3 Prevention and control of major zoonosis

3.3.1 Brucellosis

China takes comprehensive measures to prevent and control brucellosis, including zoning management, vaccination, surveillance and stamping out.

Zoning management. China practices zoning management in the prevention and control of brucellosis. The country is divided into three regions based on the proportion of counties with uncontrolled epidemic situation (sheep positive rate $\geqslant 0.5\%$ or cattle positive rate $\geqslant 1\%$ or pigs positive rate $\geqslant 2\%$) – category I region (uncontrolled counties/all counties$>$30%, including Beijing, Tianjin and other 13 provinces and Xinjiang Corps), category II region (uncontrolled counties/all counties$<$30%, 15 provinces including Jiangsu and Shanghai) and clean region (provinces with no human or animal cases, currently only Hainan).

Vaccination and surveillance. In 2014, MOA continued to guide local governments to immune against brucellosis. Voluntary vaccination was practiced in places with high prevalence rate and was filed with MOA.

Surveillance on cattle and sheep brucellosis was enhanced according to the requirements in the *national Brucellosis Surveillance Plan,* especially on breeder cattle. Various levels of animal diseases prevention and control institutions monitor all breeder cattle stations and farms within their jurisdictions, as well as on susceptible animals such as cattle and sheep in breeder farms, large-scale farms, scattered farming households, live livestock trading markets and slaughter houses. The Brucellosis specialty lab of China Animal Health and Epidemiology Center conducted etiological surveillance on breeder oxen stations across the country, and reviews local surveillance results. Brucellosis specialty lab of China Institute of Veterinary Drug Control monitored livestock in key areas.

Epidemiological investigation. According to the requirements in the *National Plan for Animal Diseases surveillance and Epidemiological Investigation 2014,* relevant parties conducted field investigation and sampling test on non-immune groups in 23 fixed spots, including Xinjiang Production and Construction Corps and 11 provinces such as Heilongjiang, Jilin and Hebei, collecting and testing 17,340 serological samples. Also, to facilitate the implementation of the *National Medium and Long Term Plan for Animal Diseases Prevention and Control*, relevant authorities conducted pilot baseline investigations for comprehensive prevention and control of brucellosis, as well as surveys on live sheep transport, health conditions and infection of brucellosis of breeder oxen. Research on risk communition strategy on brucellosis were also carried out among rural employees in pilot areas. All this provided essential data and technical support necessary to the prevention & control of brucellosis.

Emergency disposal. In 2014, brucellosis occurred in 19 provinces (municipalities or autonomous regions), namely Hebei, Shanxi, Inner Mongolia, Heilongjiang, Jiangsu, Zhejiang, Fujian, Jiangxi, Shandong, Hubei,

Hunan, Guangxi, Chongqing, Guizhou, Yunnan, Shaanxi, Gansu, Ningxia and Xinjiang, involving 28,735 cases (cattle, sheep and pigs), 41 deaths, 1,260 Slaughtered, and 23,902 destroyed. MOA instructed local animal husbandry and veterinary authorities to handle these outbreaks, and all outbreaks were controlled and eliminated in time.

3.3.2 Bovine tuberculosis

According to *Bovine Tuberculosis Surveillance Plan* issued by MOA, local governments continued to conduct surveillance to prevent and control bovine tuberculosis in 2014, and handled etiologically positive cattle in a timely manner following *Technical Standards for Prevention and Control of Bovine Tuberculosis*. Also, according to requirements in *National Plan for Animal Diseases Surveillance and Epidemiological Investigation*, epidemiological investigation of bovine tuberculosis was conducted on dairy cattle in 23 counties (cities or districts) in 11 provinces, and 2,096 whole blood samples were collected and tested. Similar investigation was conducted on breeder oxen at 29 oxen stations in 17 provinces, and 2,132 oxen were tested.

In 2014, eight provinces reported cases of bovine tuberculosis, namely Shaanxi, Xinjiang, Zhejiang, Inner Mongolia, Gansu, Sichuan, Hunan and Jiangsu, involving 783 cases, 36 deaths and 715 destroyed.

3.3.3 Schistosomiasis

In 2014, MOA continued to implement the *Key Project Plan for the Comprehensive Control of Schistosomiasiss (2009-2015)* by holding national work conference on rural schistosomiasis prevention, providing guidance on livestock examination and treatment and rural schistosomiasis control, and promoting compliance inspection. Over 1,260,000 animals were examined and treated. The MOA also summarized the work on prevention and control of schistosomiasis and rural prevention efforts since 2009, and conducted field research in seven affected provinces.

3.3.4 Echinococcosis

The Action Plan on the Prevention of Echinococcosis (2010-2015) was carried on in 2014. Pilot programs were carried out in three provinces including Qinghai and Xinjiang, as well as in Xinjiang Corps to enhance livestock surveillance and parasite expelling for dogs.

In 2014, 72 sheep/goat echinococcosis outbreaks (echinococcus granulosas infected) were reported in Xinjiang, Qinghai and Inner Mongolia, involving 266 cases, 38 deaths and 84 destroyed.

3.3.5 Rabies

MOA continued to guide local governments to vaccinate in preventing and controlling rabies, and kept on surveillance in provinces with frequent rabies occurrence following the requirements in the national *Rabies Surveillance Plan*, especially on rural dogs and cats or stray dogs and cats in cities, as well as those visiting animal hospitals. In 2014, rabies surveillance was carried out in 26 provinces (autonomous regions or municipalities) and Xinjiang Production and Construction Corps. A total of 2,333 monitoring spots were set up, and 28,085 samples were tested, including 18,624 immune antibody samples with an acceptance rate of 73.52%, and 8,119 etiological samples, in which 29 or 0.36% came out positive.

In 2014, rabies occurred in three provinces (autonomous regions), namely Inner Mongolia, Hebei and Shaanxi, involving 94 cases (dogs, cattle, sheep and camels) and 71 deaths.

3.4 Eradication of glanders and Equine infectious anaemia

Glanders has a long history in China, and Equine infectious anaemia was introduced into China in the 1950s. After the new China was established, the two diseases went prevalent in 21 and 22 provinces respectively, causing a huge economic loss to agricultural and animal husbandry production. Various levels of livestock veterinary authorities have made tremendous progress in preventing and controlling the diseases thanks to thoughtful planning, the application of technologies, and comprehensive measures combining immunization, surveillance and stamping out. In recent years, no cases of glanders or equine infectious anaemia were reported across the country, and no animals came out positive for glanders infection. Yunnan and Xinjiang were the only two provinces with a very few number of animals positive for equine infectious anaemia. In 2014, the contagious equine diseases lab (OIE reference lab for equine infectious anaemia and MOA-designated professional lab for contagious equine diseases test) of Harbin Veterinary Research Institute of the Chinese Academy of Agricultural Sciences conducted sampling surveillance on the two diseases in 23 provinces. A total of 1,879 and 4,120 samples were tested for glanders and equine infectious anaemia respectively, and all came out negative.

In December 2014, MOA held a symposium on the prevention and control of glanders and equine infectious anaemia in Kunming, Yunnan province to discuss the next steps in eradicating the two diseases. For glanders, regulation and constant surveillance are required to maintain and prove the disease-free status, and data collation should be enhanced to support the disease-free status assessment with scientific and detailed information. For equine infectious anaemia, prevention should always be prioritized and comprehensive measures following the principle of "differentiated approaches and step-by-step improvement" should be implemented. The scope and targets of surveillance should be expanded, and killing and biosafety disposal on positive animals should be strictly carried out. Quarantine supervision should be strengthened to prevent cross-region spread. Emergency preparedness should be enhanced and appropriate measures should be taken as soon as epidemic cases occur. Management of epidemics control should be enhanced by keeping good registration of equine animals and detailed records of prevention and control efforts. Besides, relevant provinces should accelerate acceptance inspections, and qualified provinces should keep up the good work in disease surveillance and data collection and collation, so as to be prepared for planned national disease-free status assessment.

3.5 Disease cleansing surveillance in breeder poultry farms and key original breeder pig farms

To implement the *National Medium and Long Term Plan for Animal Diseases Prevention and Control* and the livestock and poultry health promotion strategy, the General Office of MOA released the *Notice on Surveillance of Diseases Featuring Vertical Transmission in Breeder Poultry Farms and Key Original Breeder Pig Farms in 2014* (Nongbanyi〔2014〕No.20), requiring continued surveillance on vertically-transmitted diseases in GP or GP-above poultry farms and key original breeder pig farms across the country. It is required that four diseases–HPAI, Avian leukosis, Avian Reticuloendotheliosis and Pullorum disease–in laying hen farms handed down from the great-grandfather or grandfather generations as well as national poultry gene pools be sampled and tested by China Animal Disease Control Center, Harbin Veterinary Research Institute of the Chinese Academy of Agricultural Sciences and Shandong Agricultural University. Five diseases, namely Classical swine fever, porcine reproductive and respiratory syndrome, pseudorabies, porcine circovirus disease and porcine parvovirus disease, should be sampled and tested by China Animal Disease Control Center. Also, MOA holds regular meetings to analyze and judge the epidemic situation with scientific tools.

3.6 Prevention of exotic animal diseases

In 2014, China continued to emphasize the awareness of border animal epidemics control, enhance risk management for major exotic animal diseases, and establish national safety shield for animal epidemics in border areas under the strategic guidance provided in the *National Medium and Long Term Plan for Animal Diseases Prevention and Control (2012-2020)*. The epidemics surveillance system and emergency response mechanism for border areas were enhanced with stronger joint defense and control and better technical and material support. Risk assessment, quarantine access, overseas pre-inspection, overseas corporate registration and traceability management on entry animals and animal products were improved. Surveillance on exotic animal diseases was constantly carried out and technical training was enhanced. In 2014, the National Exotic Diseases Research Center provided four training courses on the prevention and control technologies for exotic animal diseases in Shanghai, Heilongjiang, Yunnan (Figure 3-2) and Hebei, producing over 420 grassroots technical workers and issuing over 400 promotion materials. More than 460 persons joined training in high-risk provinces and over 1,000 went to relevant provinces to attend such training courses.

Figure 3-2　Technical training course for prevention & control of exotic animal diseases (Yunnan)

3.6.1　Animal spongiform encephalopathies (BSE, scrapie)

Bovine spongiform encephalopathy (BSE) have never occurred in China. China has implemented strict prevention and control measures on BSE since 1990. First, BSE is classified as category I disease subject to compulsory notification, and great awareness promotion and training efforts guaranteed sensitivity to BSE cases of the surveillance system. Second, national BSE reference and professional labs were set up, and OIE-recognized diagnostic methods and quality standard systems were established. BSE Surveillance went on for 13 years and no cases were ever discovered. Third, import of cattle, sheep or other ruminant-derived meat and bone meal, bone meal and greaves was strictly prohibited from BSE-affected countries and regions, which effectively prevented introduction of risk commodities. Fourth, feeds regulation was enhanced. Strict prohibition was issued on the import of animal feeds from high-risk areas in 1999, and feeding ruminants with animal feeds has been fully prohibited since 2001. In September 2013, MOA officially submitted the application report to the OIE for BSE negligible risks in the Chinese mainland. After requesting further materials and two on-site inquiries at the OIE headquarters, the OIE agreed that the Chinese mainland had met the standards. At the 82nd General Session of the World Assembly of OIE Delegates in May 2014, the Chinese mainland was officially recognized by OIE as a zone with a negligible BSE risk (Figure 3-3).

Figure 3-3　Chinese Mainland was recognized by the OIE as zone with negligible BSE risk

In 2014, China kept on monitoring BSE and scrapie and testing ruminants-derived feed ingredients, and did not import live cattle and ruminants meat and bone meal from risk countries. A total of 6,463 cattle brain samples were tested for BSE throughout the year, and all came out negative with a score as high as 59,000; 3,777 sheep brain samples were tested for scrapie, and all came out negative; 1,822 batches of ruminants feeds were sampled and tested, and no ruminants-derived ingredients were discovered. Australia, New Zealand, Uruguay and Argentina were the only four countries where China imported live cattle or meat and bone meal from. The Chinese mainland maintained its status with negligible BSE risk.

3.6.2　African swine fever

African swine fever (ASF) have never occurred in China. In recent years, China has been constantly improving control on the disease to prevent its introduction. In November 2013, China joined the global technical platform for the prevention and control of African swine fever. In July 2014, MOA jointly initiated the ASF-TCP program with the FAO aimed at preventing the disease in China. The program is intended to use FAO technologies and expert resources to enhance the joint defense and control mechanism with neighboring countries, strengthen lab diagnosis technological research and reserve, and improve technical standards and contingency plans for the disease, so as to improve the overall prevention and control capacity. In 2014, surveillance on African swine fever was further enhanced. Sample collecting was expanded from Xinjiang, Jilin, Heilongjiang, Inner Mongolia and Liaoning, to Beijing, Tianjin, Shanghai and Guangdong for the first time. A total of 5,806 samples were tested throughout the year (2,501 etiological samples, in which 80 were from wild boars; 3,305 were serological samples, in which 20 were from wild boars) and all came out negative. In September 2014, veterinary and entry-exit inspection and quarantine authorities jointly participated in the emergency drill for African swine fever outbreaks in border areas in Heihe, Heilongjiang province.

3.6.3 Other exotic animal diseases

MOA continued monitoring on West Nile fever, African horse sickness, Nipah virus encephalitis, Vesicular stomatitis, Rift valley fever, Swine vesicular disease, and other exotic diseases already eradicated in China such as Rinderpest and contagious bovine pleuropneumonia, and all samples came out negative.

3.7 Prevention & control of other terrestrial animal diseases

3.7.1 H7N9 influenza

China's success in responding to H7N9 influenza since 2013 is widely recognized by the international community and all sectors of society. In 2014, MOA further enhanced such efforts when humans cases were reported in some provinces.

First, *The National Poultry H7N9 Influenza Elimination Plan* was released in time. Surveillance, epidemiological investigation and market chain analysis were employed to gain a clear picture of the time, space and herd distribution of the virus. Immediate measures were taken to eliminate the H7N9 virus in poultry in such key links as farming and market distribution, so as to lower the risk of human infection and transmission of the virus from live poultry markets to poultry farms. Implementation plans were developed by 31 provinces and Xinjiang Corps. Training programs were provided together with the FAO on the prevention and control technologies for H7N9 virus for key staff members. Supervision on H7N9 was shared with National Health and Family Planning Commission (NHFPC) and China Food and Drug Administration (SDA). Four work groups were sent to six provinces to guide local prevention and control efforts.

Second, surveillance and epidemiological investigation were conducted in live poultry markets, poultry farms and wild birds habitats, supported by a fund of RMB17.39 million. In 2014, H7N9 monitoring was conducted on animals and the environment in 9,743 live poultry markets, 19,817 poultry farms and farming households, 2,911 breeder poultry farms, 10,046 scattered farming households, 132 wild birds habitats and 298 other sites. A total of 1,139,000 samples were tested, including 784,000 serological samples and 355,000 etiological samples, and 61 or 0.02% came out H7N9 positive. Measures were taken on positive herds according to the requirements in *Emergency Disposal Guide for Animal H7N9 Influenza*.

Third, regulation on distribution and transport was enhanced to limit transferring poultry from high- to low-risk areas. The report-on-arrival and quarantine observation systems were strictly carried out for breeder poultry, young breeder poultry and farming poultry transported across provinces.

Fourth, bio-safety management was strengthened in live poultry trading markets and poultry farms. Mandatory market-off disinfection was carried out regularly, animal farms were subject to animal epidemic prevention conditions censorship, farming management and prevention facilities building were improved, and the establishment of compartments was encouraged. All this were intended to improve bio-safety in poultry farms.

3.7.2 Peste des petits ruminants

In 2014, 251 outbreaks of peste des petits ruminants were reported in China, involving 33,041 cases, 14,681 deaths and 51,333 slaughtered and destroyed. MOA was in charge of the prevention and control efforts. A total of RMB 48,620,000 was allocated for the disposal of emergent cases, and immunization was carried out in high-risk areas. *The Notice on Improving Regulation over Cross-province Live Sheep Transport* was released for better control of live sheep transport; outbreak investigation was conducted and emergency disposal measures were taken on sick and exposed sheep; surveillance was enhanced, given that 35,387 samples were tested in 2014 (17,744 etiological samples and 17,643 serological samples); training programs and promotion events were provided for over 4,000 persons in the forms of expert lecture and advertising video; several supervisory teams were sent to various locations to guide local quarantine, killing, biosafety disposal and immunization efforts. As a result, outbreaks were soon contained without further spread, and the steady growth of the industry and mutton supply were guaranteed.

3.7.3 Porcine epizootic diarrhea

In 2014, porcine epizootic diarrhea occurred in 26 provinces and Xinjiang Production and Construction Corps, involving 228,782 sick and 34,378 dead pigs.

In 2014, MOA continued guiding local governments to prevent and control common and frequent diseases, including porcine epizootic diarrhea. Under *The Guiding Opinions on Prevention and Control Technologies for Porcine Epizootic Diarrhea* released by MOA, local governments improved prevention and control over the disease, strengthened technical guidance and services, and implemented a series of comprehensive measures, for example, vaccine immunization, disinfection, ventilation and temperature control in delivery rooms, timely assessment of sow health conditions and specific measures to improve daily farming management, better management over piglet farming, strict implementation of disinfection and introduction quarantine measures, and strengthened disease cleansing and biosafety disposal.

3.7.4 Other OIE listed diseases

See Table 3-6 for details.

Table 3-6 Other OIE listed diseases that occurred in China, 2014

Disease	Outbreaks	Species	Cases	Deaths	Slaughtered	Destroyed
Anthrax	20	Cattle, pig	113	80	0	26
Pseudorabies	144	Pig	2,960	785	0	1,388
Paratuberculosis	10	Cattle	437	0	0	437
Trichinosis	2	Pig	17	1	0	0
Epidemic encephalitis B	62	Pig	412	79	55	83
Bovine babesiosis	5	Cattle	11	1	0	1
Bovine hemorrhagic septicemia	66	Cattle	1,249	275	13	694
Infectious bovine rhinotracheitis/infectious pustular vaginitis	12	Cattle	27	4	0	14
Trichomoniasis	5	Cattle	252	0	0	0
Trypanosomosis	31	Cattle	656	12	0	275
Bovine virus diarrhea	64	Cattle	536	91	0	110
Goat pox/sheep pox	200	Goat/sheep	7,873	442	143	1,163
Sheep epididymitis	*	Goat/sheep	*	*	*	*
Goat arthritis/encephalitis	13	Goat/sheep	104	8	0	22
Goat contagious pleuropneumonia	225	Goat/sheep	4,279	1,023	25	2,022
Enzootic abortion of ewe	22	Goat/sheep	167	0	0	0
Salmonellosis	90	Sheep	789	115	0	242
Porcine cysticercosis	15	Pig	180	0	4	7
Swine transmissible gastroenteritis	4,597	Pig	189,310	13,873	957	35,433
Avian infectious bronchitis	2,482	Poultry	413,540	18,888	1,085	35,802
Avian infectious laryngotracheitis	1,388	Poultry	310,375	16,138	443	33,976
Duck virus hepatitis	355	Poultry	35,799	4,897	246	9,904
Avian typhoid	260	Poultry	26,732	2,753	29	2,483
Infectious bursal disease	698	Poultry	222,110	15,243	270	31,240
Avian mycoplasmosis	175	Poultry	34,602	555	158	909
Avian chlamydia	10	Poultry	5,151	2	0	2
Pullorum idsease	8,595	Poultry	1,250,871	74,843	4,941	121,157
Low pathogenic avian influenza	60	Poultry	127	0	0	807,295
Rabbit viral hemorrhagic disease	138	Rabbit	3,639	2113	28	986

3.8 Aquatic animal disease prevention and control

In 2014, aquatic animal diseases reported in China included spring viraemia of carp (SVC), infectious haematopoietic necrosis (IHN), Taura syndrome, white spot syndrome (WSS), yellow head disease, infectious hypodermal and haematopoietic necrosis (IHHN), white tail disease, etc.

In 2014, MOA implemented the *National Surveillance Plan for Aquatic Animal Diseases* which listed five major aquatic animal diseases to be specially monitored, namely spring viraemia of carp (SVC), white spot syndrome (WSS), infectious haematopoietic necrosis (IHN), Koi hepesvirus disease (KHVD) and Cryptocaryon irritans.

3.8.1 Spring viraemia of carp (SVC)

Surveillance on SVC in cyprinid fish was carried out in 18 provinces / autonomous regions/municipalities including Beijing. A total of 891 batches of samples were collected in 2014, and 22 batches were found to include SVC positive samples with an average positive rate of 2.5% and including the species of carp, crucian carp, fancy carp, goldfish, silver carp, jian carp, bighead carp, etc. No epidemics of SVC occurred in the actual aquatic cultivation and production in 2014.

3.8.2 White spot syndrome (WSS)

Surveillance on WSS in crustacea aquatic animals was carried out in 9 provinces/autonomous regions/municipalities including Tianjin. A total of 1152 batches of samples were collected in 2014, among which 191 batches were found to include WSS positive samples with an average positive rate of 16.6%, covering the species of litopenaeus vannamei, procambarus clarkii, Fenneropenaeus chinensis, Macrobrachium rosenbergii, penaeus japonicus, freshwater shrimp and crab, etc. Epidemics of WSD in occulitopenaeus vannamei and procambarus clarkii occurred in some regions in 2014.

3.8.3　Infectious haematopoietic necrosis (IHN)

Surveillance on IHN in salmons was carried out in five provinces/municipalities, namely Beijing, Hebei, Liaoning, Shandong and Gansu Province. A total of 298 batches of samples were collected in 2014, among which 61 batches were found to include IHN positive samples with an average positive rate of 20.5% and mainly covering the species of rainbow trout and golden trout. Epidemics of IHN occurred in some regions in 2014.

3.8.4　Koi hepes-virus disease (KHVD)

Surveillance on KHVD was carried out in 14 provinces/autonomous regions/municipalities for the first time, and the monitored species were carp and fancy carp. A total of 318 batches of samples were collected in 2014, and 4 batches were found to include KHVD positive samples with an average positive rate of 1.3%. No epidemics of KHVD occurred in actual aquatic cultivation and production.

3.8.5　Cryptocaryon irritans

Surveillance on cryptocaryon irritans in marine fish was carried out in 3 provinces (Zhejiang, Fujian and Guangdong Province). A total of 584 batches of samples were collected in 2014, among which 96 batches were found to include positive samples with an average positive rate of 16.4% and covering the species of large yellow croaker, black sea bream, seabream gilthead, Plectorhinchus cinctus, Rachycentron canadum, Trachinotus ovatus and sciaenops ocellatus in various sizes. Epidemics of cryptocaryon irritans occurred in some regions in 2014.

3.9 Development of the mechanism of animal disease prevention & control

3.9.1 Continuously improving emergency mechanism

In 2014, MOA and governments at all levels continued to improve the contingency mechanism and plan for major animal diseases, emergency personnel reserve, emergency resource reserve, drills and training for emergency to beef up emergency response. All local authorities formulated contingency plans for major animal epidemics in their own administrative regions, with a contingency plan system covering the whole country initially established. In 2014, drills for emergency at provincial, municipal and county levels were organized to cope with major animal diseases in Tianjin, Hebei, Shandong, Heilongjiang, Jilin, Qinghai, Guizhou, Sichuan, Hunan, Guangxi, Guangdong, etc (Figure 3-4). Besides, in response to flood, mudslide and landslide that happened in some regions, super typhoon Ramasoon that inflicted heavy losses on several provinces including Hainan, Guangdong, Guangxi and Hunan, and strong earthquakes that happened in a row in Kangding County in Sichuan Province, Gudian County and Gujing County in Yunnan Province, MOA took prompt actions, made timely arrangement and deployment, dispatched multiple batches of emergency personnel for front-line guidance and supervision of emergency disposal, and prevented animal epidemics from occurring.

Figure 3-4　Emergency drill for animal epidemics in Qinghai in 2014

3.9.2　Continuously improving designated liaison system for major animal disease prevention & control

In 2014, *MOA* issued *Notice on the Adjustment of the Working Group of Designated Liaison System for Major Animal Disease Prevention and Control* (Nongbanyi〔2014〕NO.44), with an aim to intensify major animal disease prevention and control, regional joint prevention and control, and the working group of designated liaison system.

The working group of the designated liaison system was made up of disease prevention & control group and vaccines supervision group. The disease prevention and control group consisted of 6 smaller groups responsible for investigating the situation of major animal disease prevention & control in designated provinces, supervising and encouraging the implementation of relevant policies and measures, guiding the stamping out of epidemics, and coordinating joint prevention and control of diseases. The vaccines supervision group was in charge of the coordination of emergency production and supply of vaccines for animal epidemics such as avian influenza and foot-and-mouth disease, the supervision and inspection of the production and quality of vaccines produced in designated regions, the investigation of the usage of vaccines when epidemics occur and the reasons for immunity failure, the assessment of local immune procedure, etc. In 2014, a total of 617 person-times of 141 supervision groups were dispatched by MOA to provide guidance for regional animal disease prevention and control.

3.9.3 Continued zoning management

In 2014, MOA continued zoning management to control animal diseases, proactively explored zoning management modes, improved rules and standards governing disease-free zones and compartments, and accelerated the establishment of disease-free zones and compartments.

First, relevant rules and technical norms governing disease-free zones and compartments were revised and improved. In 2014, specialists of relevant authorities revised the *Measures for Administration of Specific Animal Disease Free Zones*, *Technical Norms of the Administration of Specific Animal Disease Free Zone* and *Norms of Specific Animal Disease Free Compartment*.

Second, surveillance and supervision were intensified to keep the disease-free state of disease-free zones. Conforming to *2014 Plan of National Animal Disease Surveillance and Epidemiological Investigation*, monitoring continued in three foot-and-month disease free zones, which were Hainan, Liaoning Province and Yongji in Jilin Province, as well as the disease-free zone of Guangzhou Province and adjacent horse activity areas. In November 2014, supervision was carried out in three disease-free zones, namely Hainan Province, Conghua in Guangdong Province and Yongji in Jilin Province.

Chapter 4

Veterinary Administrative Law Enforcement

The Chinese government has constantly strengthened veterinary administrative law enforcement, animal quarantine and animal products at the production places and slaughterhouse, review of animal epidemic prevention conditions, animal health supervision and law enforcement, supervision of slaughtering, long-term mechanism of the bio-safety disposal of animals dying of diseases, and bio-safety supervision of veterinary laboratory, carried out cleaning-up and rectification on medical institutions, and improved animal identification and animal product traceability system, constantly improving the ability of animal health and animal product quality and safety supervision.

4.1 Animal Health Supervision

4.1.1 Inspection of animals and animal products

Inspection of animal and animal products is administrative licensing behavior of animal health supervision authorities according to *Law of the P. R. China* On *Animal Epidemic Prevention*. Local animal health supervision authorities performed inspection on animals and animal products in accordance with *Animal Epidemic Prevention Law of the P. R. China, Adminis-*

trative Measures for Animal Inspection, and standard procedures of animal quarantine; issued Animal Quarantine Conformity Certificates if quarantine inspection is passed. In 2014, MOA continued to carry out uniform animal inspection job logging nationwide, conducted traceable management of animal inspection, made animal inspection process strictly conducted and standardized the issuing of Animal Quarantine Conformity Certificates. Inspection at the production place covered 11.595 billion birds and animals (including 473 million live pigs, 25 million heads of cattle, 77 million heads of sheep, 10.974 billion birds and 46 million heads of other animals), among which over 2.96 million birds and animals were detected with diseases. Inspection at the slaughterhouse covered 6.367 billion birds and animals (including 369 million live pigs, 50 million heads of cattle and sheep, 5.934 billion birds and 14 million heads of other animals), among which over 5.9 million birds and animals were detected with diseases.

4.1.2　Review of animal prevention epidemic conditions

Local veterinary authorities act according to *Law of the P. R. China On Animal Epidemic Prevention* and *Review Measures for Animal Epidemic Prevention Conditions*, conduct review on animal epidemic prevention conditions of animal farms (animal raising areas), isolation sites, animal slaughtering and processing sites, and animal and animal product bio-safety disposal sites within their respective administrative areas, and issue the Certificate for Qualified Animal Disease Prevention Conditions. In 2014, altogether 666,000 Certificates for Qualified Animal Disease Prevention Conditions were issued across the country. Animal health supervision authorities of all levels are responsible for daily supervision and law enforcement of the above-mentioned sites that have qualified animal epidemic prevention conditions within their respective administrative areas.

4.1.3　Animal Health Supervision and Law Enforcement

Firstly, work style in animal health supervision was perfected. MOA further carried out nationwide movement of work style standardization in animal health supervision and law enforcement, launched "the movement of ability enhancement in animal health supervision", strengthened law enforcement team, standardized the administration of quarantine certificate and badging, and comprehensively improved the quality and professional ability of personnel in animal health supervision authorities of various levels.

Secondly, behaviors in animal health supervision and law enforcement was standardized. Ten inspection teams were sent to 16 provinces to carry out special programs of animal quarantine supervision and law enforcement, examine implementation of relevant laws and regulations as well as the "Six Bans", and the standardization of issuing quarantine certificates, the handling of cases in animal health supervision and law enforcement, and management of animal health badging. 13 typical cases violating law and regulations in animal quarantine supervision and law enforcement that happened in recent years were together notified, which contributed to the interpretation of laws with real cases, in order to intensify supervision and further standardize behaviors in animal quarantine and animal health supervision and law enforcement. In 2014, animal health supervision authorities of various levels performed supervision and inspection on 89,300 markets of poultry and livestock, 30,000 slaughterhouses, and 1.4 million large-scale farms breeding poultry and livestock. Supervision in circulation links covered 0.38 billion animals, 2.1 billion birds and 6.5353 million tons of animal products. Moreover, nearly 30,000 cases violating *Law of P. R. China on Animal Epidemic Prevention* were investigated and penalized, which guaranteed the healthy development of animal husbandry and the quality and safety of animal products.

Thirdly, informatization management of animal health supervision and law enforcement was promoted. The software platforms that issue electronic animal quarantine certificates were developed at both national and provincial levels, and Jiangsu, Jiangxi, Xinjiang Production and Construction Corps became the pilot places for issuing electronic animal quarantine certificates.

Fourthly, the training of animal health supervision was enhanced. In August 2014, the third advanced training course on animal health supervision was organized in Hulun Buir of Inner Mongolia Autonomous Region. In this course, experts of relevant fields were invited to provide training and have exchanges on administrative legislation and administrative law enforcement, agricultural informatization and the application of cutting-edge informatization technologies, technological innovation and agricultural development, and risk analysis and prevention strategies for African swine fever. Training on animal health supervision was organized in Hubei Province, Inner Mongolia Autonomous Region and other regions, and training on animal health supervision information statistics was organized in Kunming of Yunnan Province.

4.1.4　Bio-safety disposal of animals dying of diseases

In 2014, MOA continued to improve the mechanism of bio-safety disposal of animals that die of diseases and strengthen the supervision of bio-safety disposal of animals that die of diseases. In 2014, bio-safety disposal was conducted on more than 20 million heads of pigs that died of diseases in the link of farming in large-scale farms.

Firstly, the establishment of long-term mechanism of bio-safety disposal of animals dying of diseases was promoted. In 2013, pilot programs of long-term mechanism of bio-safety disposal of pigs dying of diseases were launched by MOA in 212 counties/cities/autonomous regions in 19 provinces. Each region proactively explored and gradually established long-term mechanism of bio-safety disposal of pigs dying of diseases, well-developed system of the collection of pigs dying of diseases and appropriate bio-safety disposal methods, established bio-safety disposal mechanism and insurance linkage. In 2014, the State Council issued the *Instruction on the Establishment of the Mechanism of Bio-safety Disposal of Animals Dying of Diseases* (The State Council Document〔2014〕No.47), which put forward the general requirement of promoting ecological civilization, with aims of timely treatment, being clean and environmentally friendly and rational utilization; the principles put forward included combining overall planning and responsibility in administrative areas, combining government supervision and market operation, combining financial subsidy and insurance linkage, and combining centralized, voluntary treatment; with the ultimate aim to establish a scientific, well-developed and highly-efficient mechanism of bio-safety disposal of animals dying of diseases which covers farming, slaughtering, operation and transportation. At the same time, MOA actively promoted the publicity and implementation by organizing video and telephone conferences, compiling *100 Questions about Bio-safety Disposal of Animals Dying of Diseases*, printing publicity posters about bio-safety disposal, carrying out theme publicity of bio-safety disposal of animals that die of diseases, and publicizing the laws, regulations and policies of bio-safety disposal, in order to promote the establishment of bio-safety disposal mechanism of animals dying of diseases.

Secondly, inspection was carried out to ensure the responsibility in each administrative region be fulfilled. In order to strengthen the supervision and administration on bio-safety disposal of animals dying of diseases, the *Emergent Notice of the General Office of MOA on Strengthening Supervision and Administration on Bio-safety Disposal of Animals Dying of Diseases* (Announcement of Nongbanyi〔2014〕No.9) was issued. Inspection on the bio-safety disposal of animals that die of diseases was organized twice, and 13 inspection groups were sent separately to 25 provinces, where they paid visits to large-scale live pig farms, sites of bio-safety disposal, collection sites and so on to investigate the implementation status of long-term mechanism of bio-safety disposal of pigs that die of diseases and the progress of bio-safety disposal in pilot places.

Thirdly, response mechanism on the events of discarding animals dying of diseases was established. Positive responses were made on the events of floating pigs that happened in a row in Jiangxi Province, Fujian Province, Qinghai Province, Shandong Province and Jiangsu Province, and local authorities were supervised and urged to make timely and proper treatment and make relevant information release well. On-site guidance and supervision on bio-safety disposal of floating dead pigs were organized in Jiangxi Province.

4.1.5 Clean-up and Rectification of Animal Medical Institutions

MOA carried out a cleaning-up and rectification among animal medical institutions nationwide for three months to standardize animal disease diagnosis, treatment and veterinarian practices and improve service level of animal medical institutions and licensed veterinarians. In the campaign, 410 illegal cases were investigated and penalized, with 124 animal hospitals and 1096 animal clinics withdrew and closed, and 486 veterinary licenses were withdrawn. This campaign clamped down on illegal practices of animal medical institutions and licensed veterinarians, banned illegal institutions according to law, standardizing the administration of animal clinics and hospitals and bettering the market order of animal disease diagnosis and treatment.

4.1.6 Animal identification and animal disease traceability system

On the basis of animal identification, farming record and epidemic prevention record, the system realizes traceable animal disease supervision by means of mobile smart reading devices, which enables a series of functions including information collection in the processes of immune injection, inspection at the production place, transportation supervision and inspection at the slaughterhouse, as well as network transmission, computer analysis and processing, and the query and output by mobile smart reading devices. In 2014, MOA continued to promote animal identification and animal disease traceability system by improving relevant infrastructure, identification management and data transmission, in order to level up standardization and informatization of the system.

Firstly, relevant infrastructure was beefed up. Central database of the system was created and put online for trial operation, which greatly reinforced technical support and service ability of the system.

Secondly, evaluation indexes of the system were constantly improved. Evaluation indexes, contents and evaluation methods of the system were clarified in the aspects of institutional framework, management of ear tag and facilities, information collection and transmission, personnel training, etc.; thus, effect of the system on the traceability of major animal diseases, supervision of animal product quality and safety were constantly improved.

Thirdly, ear tag management and information transmission were constantly promoted. In 2014, a total of 739 million ear tags were applied, and 151 million pieces of information were uploaded to the central database. Trials of wearing electronic identification and information transmission were launched in Liaoning province, Inner Mongolia Autonomous Region, and Xinjiang Autonomous Region, based on the application of their own information-based platforms.

Fourthly, training on system operation was actively carried out. Training was organized on informatized knowledge, such as product quality control, traceability operation and "internet+" for 4 times, attended by 260 trainees including the manufacturers of identification and reading devices and excellent personnel concerned with the system. Training on system operation was organized for more than 50 times at various levels with nearly 10,000 trainees attending.

4.2 Bio-safety Supervision of Veterinary Laboratories

In 2014, MOA continued to strengthen bio-safety management of pathogenic microbe laboratories. Seminars were organized on bio-safety management of pathogenic microbe laboratories, with bacterial (viral) strains preservation, bio-safety measures and other relevant issues as research topics. Training courses were organized on veterinary system laboratory quality control and bio-safety, with laboratory quality control and bio-safety management measures as the major content of discussion and training. The examination and verification on bio-safety of highly pathogenic microbe activities were rigorously conducted, with highly pathogenic animal microbe laboratory activities in South China Agricultural University and Yangzhou University approved. The accuracy of disease detection was examined in 31 provincial veterinary laboratories and the veterinary laboratory of Xinjiang Production and Construction Corps, the accuracy rates of detection of foot-and-mouth disease, avian leucosis, H7N9 sub-type influenza, and porcine reproductive and respiratory syndrome were 100%, 100%, 93.8% and 96.9% respectively. The research of veterinary laboratory layout planning was intensified by drawing on advanced experience about network building in foreign laboratories and the administration measures of OIE Reference Centers were taken as references for veterinary laboratory administration in China.

Chapter 5
Veterinary Drug Production and Supervision

Chinese veterinary drug laws & regulations, technical standards and specifications are basically complete, veterinary drug products safety level has significantly increased, the soundness of veterinary drug regulatory system has been further improved, the supervision ability and level has been enhanced, and veterinary drug industry is in healthy development. In 2014, MOA continued to improve management system, optimize working mechanism, strengthen veterinary drug quality supervision and law enforcement, further standardize veterinary drug production, management and utilization, thus improving veterinary drug quality.

5.1 Production, Research & Development of Veterinary Drugs

By the end of 2014, there was a total of 1,809 veterinary drug production enterprises nationwide, including 82 biological production enterprises, with about RMB 44 billion of output value, and RMB 40.7 billion sales volume, totally recruiting 165,000 employees. Veterinary drug industry was expanding gradually, with increase of production and sales year by year.

Veterinary biological product enterprises totally realized output value of RMB11.437 billion, products mainly include livestock and poultry live vaccine and inactivated vaccines. In 2014, national live vaccine production capacity was about 452.5 billion plume (head), inactivated vaccine production capacity of about 59 billion ml. Veterinary chemicals enterprises realized the total output value of RMB32.518 billion, with products mainly including antimicrobial drug, anti-parasite drug, and anti-pyretic analgesic antiinflammatory drug. Antimicrobial drug production capacity was 116,000 tons, anti-parasite drug production capacity was 9,500 tons, and anti-pyretic analgesic anti-inflammatory drug 0.28 tons. Production enterprises were still mainly small businesses and small and medium enterprises, with small (annual sales under RMB5 million) and medium-sized enterprises accounting for 37.5% and 51% respectively, micro enterprises and large enterprises (annual sales above RMB200 million) accounting for 8.5% and 3% of the total number of enterprises respectively.

In 2014, veterinary drug production enterprises invested a total of RMB2.728 billion in R&D, among which RMB0.843 billion investment in biological drug enterprises, and RMB 1.885 billion research & development capital investment of chemical drug enterprises. In 2014, MOA issued a total of 20 biological product certificates for new veterinary drug, among which 1 for class Ⅰ, 4 for class Ⅱ, 15 for class Ⅲ, involving foot-and-mouth disease, Newcastle disease and nearly 20 kinds of animal diseases; issued a total of 34 chemical certificates for new veterinary drug, among which 8 for class Ⅱ, 13 for class Ⅲ, 3 for class Ⅳ, and 10 for class Ⅴ.

5.2 Supervision of Veterinary Drug Production

5.2.1 Veterinary Drug GMP Management

In 2002, MOA issued "*Veterinary Drug Production Quality Manage-*

ment Specifications" (Veterinary Drug GMP), started to implement the veterinary drug GMP system on veterinary drug enterprise, and established the veterinary drug industry entry and exit mechanism. In recent years, MOA further perfected veterinary drug GMP management, revised management methods for implementing veterinary drug GMP and evaluation criteria of examination and acceptance, carried out veterinary drug GMP inspector training, supervision and inspection, strengthened GMP inspector team building, and continued to send inspection teams to conduct site inspection and acceptance for relevant enterprises. In 2014, a total of 295 veterinary drug production enterprises were issued with "Veterinary Drug GMP Certificate" and "Veterinary Drug Production License".

5.2.2 Veterinary Drug Products QR Code Traceability Pilot

Establish and complete national veterinary drug product traceability information system, implement veterinary drug products "QR Code" identification, set up unified national traceability system, and realize traceability management on the production, management and utilization of veterinary drug products. In February 2014, national veterinary drug enterprises products QR code tracing system kick-off meeting was held in Luoyang, Henan province, totally 108 enterprises (including groups) participated in traceability system production pilot application work, 4 operating enterprises and supervision and law enforcement and veterinary administrative departments of 2 provinces and autonomous regions took part in management and supervision function tests, 13 equipment businesses (including automation integrators) and 52 printing enterprises participated in pilot work. By the end of 2014, the traceability system pilot work went on smoothly, with smooth running of system, laying foundation for full implementation of veterinary drug quality safety traceability management in China.

5.2.3 Major Animal Disease Vaccine Quality Supervision

In 2014, MOA strengthened major animal disease vaccine regulation, to ensure the quality and supply of vaccine. Continue to work on the production, supply and quality supervision on avian flu and other major animal disease vaccine, implement prevention and control in spring and autumn, flight checks, supervision sampling inspection and other measures, to ensure safety and effectiveness of major animal disease vaccine. In 2014, major animal disease vaccine batch issuance rate and anti-counterfeiting label pasting rate reached 100%, completed supervision and sampling inspection on a total of 333 batches of veterinary biological products, with 326 batches qualified (pass rate 97.9%).

5.2.4 Veterinary Drug Quality Supervision Sampling Inspection

In 2014, MOA continued to implement veterinary drug quality supervision sampling inspection plan, strengthened sampling result utilization, organized and carried out counterfeited veterinary drug investigation and treatment activities, and implemented in-depth "investigation and cracking down" united actions. In January 2014, MOA issued "*2014 Veterinary Drug Quality Supervision Sampling Inspection Plan*", strengthening sampling inspection proportion on key links, key products and key enterprises, to collect related products from veterinary drug production, management and utilization links for the test. In 2014, a total of 15,124 batches of veterinary drug went through sampling inspection, with 14,415 batches qualified (pass rate 95.3%), and year-on-year increase of 2.1%. Among them, total sampling in the production link were 2,590 batches, with 2,521 batches qualified (pass rate 97.3%) (Table 5-1). MOA released veterinary drug supervision sampling inspection report on a quarterly basis, to perfect veterinary drug quality tracking inspection system, establish and enhance close communication and coordination between departments and regions, as well as united supervision and other mechanisms, strengthen supervision over veterinary drug sampling and testing, personnel training, standardize supervision and sampling inspection behavior, investigate and handle counterfeited veterinary drug according to laws, with heavy punishment, increase intensity of exposure to illegal veterinary drug crimes, and select typical cases to perform impact propaganda and report, to carry out warning education based on cases.

Table 5-1　2014 National Veterinary Drug Quality Supervision Sampling Inspection Situation

		Quarter 1	Quarter 2	Quarter 3	Quarter 4
Identification Sampling Inspection	Sampling Inspection Quantity (Batch)	179	338	416	497
	Qualified Quantity (Batch)	173	320	382	470
	Pass rate (%)	96.6	94.7	91.8	94.6
Supervision Sampling Inspection	Sampling Inspection Quantity (Batch)	1814	2866	2828	3342
	Qualified Quantity (Batch)	1745	2375	2695	3162
	Pass rate (%)	96.2	95.4	95.3	94.6
Tracking Sampling Inspection	Sampling Inspection Quantity (Batch)	405	737	605	684
	Qualified Quantity (Batch)	384	713	585	666
	Pass rate (%)	94.8	96.7	96.7	97.4
Directional Sampling Inspection	Sampling Inspection Quantity (Batch)	68	115	135	95
	Qualified Quantity (Batch)	66	111	121	87
	Pass rate (%)	97.1	96.5	89.6	91.6
Total	Sampling Inspection Quantity (Batch)	2466	4056	3984	4618
	Qualified Quantity (Batch)	2368	3879	3783	4385
	Pass rate (%)	96.0	95.6	95.0	94.9

5.3 Supervision in Veterinary Drug Operation

To strengthen quality management of veterinary drug operation, and guarantee veterinary drug quality, on the basis of pilot tests, compulsory veterinary drug business access system was enforced since 2009. In 2010, MOA promulgated *"Veterinary Drug Good Supply Practice"*(GSP), and implemented whole process quality control on procurement, storage, sales and other veterinary drug operation links. In recent years, the entire country pushed implementation of veterinary drug GSP with full strength, carried out veterinary drug operations sorting and standardization, strengthened veterinary drug operations law enforcement inspection, and enhanced veterinary drug sampling inspection in business operation. In 2014, totally 10,147 batches of veterinary drug had sampling inspection in business operation, with 9634 batches qualified (pass rate 94.9%).

5.4 Supervision in Veterinary Drug Utilization

5.4.1 Setting Up Prescription Drug Management System

In February 2014, MOA issued *"Rural Vet Basic Drug Directory"*; from March 1st, started to implement the updated *"Veterinary Prescription and OTC Drug Management Method"*, *"Veterinary Prescription Drug Varieties Directory (First Batch)"* and *"Veterinary Drug Products Specification Template"*. Since then, the Veterinary Bureau of MOA issued the notice of implementation, and put forward requirements from the implementation significance, publicity training, preparation, feedback and other aspects. In April, the Veterinary Bureau of MOA held the "training workshop on '*Measures for the Administration of Veterinary Prescription and OTC Drug*' and veterinary drug traceability system building" in Harbin city in Heilongjiang province, carried out discussion and training on the veterinary drug classification management and veterinary drug traceability system building. etc. Strengthen management on veterinary prescription drug, enhance daily regulation of veterinary drug operation market, strictly control veterinary drug operators to sell veterinary prescription drug according to veterinary prescription, and standardize aquaculture drug use.

5.4.2 Strengthening Veterinary Drug Residue Monitoring

MOA formulated and implemented "*2014 Animals and Animal Products Veterinary Drug Residue Monitoring Plan*", continued to carry out veterinary drug residue test, enlarge sampling coverage, increase sampling frequency, and implement traceability on positive samples (Table 5-2, Table 5-3). The tested livestock and poultry animal tissues include chicken, chicken liver, eggs, beef, milk, lamb, pork, pork liver, pig urine, a total of nine, and the tested drug and harmful chemicals include diethylstilbestrol, chloromycetin, metronidazole/ metronidazole and its metabolites, nitrofurans metabolites, β-receptor agonist, assimilation hormone, kappa oxygen residual marker, olaquindox residue marker, fluoroquinolones, sulfonamides, tetracyclines, β-lactam, aminoglycosides, lincosamide, avermectin, tylosin, tilmicosin, ceftiofur, dexamethasone, clopidol, diclazuril, nicarbazin residual marker, thiamphenicol, macrolides and totally 24 varieties (classes). The samples covered 30 provinces (autonomous regions and municipalities) except Tibet. In 2014, a total of 13,164 batches of livestock and poultry animals and their products had testing on the veterinary drug residue, with pass rate 99.96%; unqualified samples were 2 batches of pork products and 3 batches of chicken products, with detected excess materials of sulfadimidine and chloramphenicol respectively. In addition, bee products residue test was also carried out in 2014, totally testing 250 batches of bee products, and tested drug included chloramphenicol, sulfonamides, fluoroquinolones, nitroimidazole, nitrofurans metabolites, tetracyclines and a total of 6 varieties (classes), with pass rate of 94.4%, 14 batches of samples exceeding standard.

Table 5-2 2014 Test Content of Veterinary Drug Residues in Animal Products

Animal	Organs	Quantity (Batch)	Drug Residue
Chicken	Egg	526	fluoroquinolones
	Chicken Liver	677	Sulfonamides, chloramphenicol, dimetridazole/metronidazole
	Chicken	3236	Diclazuril, fluoroquinolones, sulfonamides, diethylstilbestrol, chloramphenicol, clopidol, nicarbazin residual marker, tetracycline, tylosin, tilmicosin, nitrofurans metabolites
Cow	Beef	942	Abamectin, clenbuterol, assimilation hormone, ceftiofur
	Milk	2690	β-lactam, avermectin, aminoglycoside, dexamethasone, fluoroquinolones, sulfonamides, thiamphenicol, lincosamide and macrolides, chloramphenicol
Sheep	Mutton	258	Sulfonamides, chloramphenicol
	Pork Liver	621	β- receptor agonist, kappa oxygen olaquindox residue marker
Swine	Swine Urine	265	β- receptor agonist
	Pork	3949	Dimetridazole/metronidazole, dexamethasone, sulfonamides, tetracycline, ceftiofur, fluoroquinolones, telmisartan, nitrofurans metabolites

Table 5-3 2014 Veterinary Drug Residue and Compounds Test in Animal Products

Drug and Compounds	Quantity (Batch)
Fluoroquinolones	2391
Sulfonamides	2324
β-laceam	1027
Chloramphenicol	975

(continued)

Drug and Compounds	Quantity (Batch)
Tetracyclines	962
Nitrofurans metabolites	927
β-receptor agonist (including 200 batches of clenbuterol single variety)	891
Dexamethasone	728
Ceftiofur	566
Avermectins	428
Tilmicosin	405
Dimetridazole /metronidazole and its metabolites	211
Kappa oxygen olaquindox residue marker	195
Clopidol	162
Tylosin	161
Thiamphenicol	160
Assimilation hormone	151
Minoglycosides	100
Diclazuril	100
Diethylstilbestrol	100
lincosamide and macrolides	100
Nicarbazin residual marker	100
Total	**13164**

5.4.3 Special Rectification on Veterinary Antimicrobial Drug

In February 2014, MOA issue "*2014 Agricultural Products Quality Safety Special Rectification Plan*", organized and carried out the 7 special rectification actions including the "Special Rectification Action on Veterinary Drug", with focus on enhancing supervision and regulation on farms (community, household) drug use safety, and cracking down on illegal drug abuse behavior. In the production link, focus on strengthening supervision and regulation on illegal production without conforming to national standards for veterinary drug, especially for unauthorized changes to formula, illegal addition of banned veterinary drug, illegal behavior on human drug or others, intensify crackdown efforts; in business operation, focus on standardization of operation activities on veterinary antibiotic drug, especially veterinary prescription drug, strengthen investigation and treatment on counterfeited veterinary antibiotic drug, strictly investigate unauthorized sales of veterinary prescription drug without the veterinary prescription, carry out sorting and rectification of veterinary drug label instructions; in utilization link, focus on strengthening supervision on veterinary antibiotic drug use, especially veterinary prescription drug use, intensify supervision and guidance as well as inspection, severely crack down on drug use of ultra-dose and beyond scope, illegal use of active pharmaceutical ingredients, not performing withdrawal period, use of veterinary prescription drug without veterinary prescription and other illegal behaviors.

5.4.4 Surveillance on Bacterial Drug Resistance

In 2014, MOA continued to organize implementation of bacterial drug resistance surveillance on products of animal origin, to find out current status of bacterial drug resistance and changes, forecast the trend of changes in bacterial drug resistance, to provide scientific basis for the formulation of veterinary drug management policy and standardized, rational use of veterinary drug.

In 2014, China Institute of Veterinary Drug Control, China Animal Health and Epidemiology Center and other relevant units collected a total of 12,247 samples from relevant provinces throughout the country, carried out isolation identification on Escherichia coli, enterococcus, salmonella, staphylococcus aureus and campylobacter jejuni, etc, and tested resistance of these bacteria for 8-13 kinds of antimicrobial drugs, analyzed the Minimal Inhibitory Concentration (MIC) and drug resistance characteristics.

5.5 Veterinary Appliances Supervision

In recent years, MOA has constantly strengthened regulation on veterinary apparatus and instruments, actively promoted veterinary instruments legislation, continuously developed veterinary appliances safety census, steadily promoted veterinary appliances supervision, sampling inspection and risk assessment, irregularly carried out supervision inspection on livestock QR code ear tags, etc. In 2014, veterinary appliances supervision sampling inspection was carried out on veterinary continuous syringe, veterinary metallic syringe, veterinary injection needle, veterinary transportation freezer and a total of four varieties of 207 samples, including 46 veterinary continuous syringes, 41 veterinary metallic syringes, 99 veterinary injection needle and 21 veterinary transportation freezers, with inspection pass rate of 71.74%, 68.29%, 69.70% and 19.05% respectively, and overall pass rate was 64.73%.

Chapter 6
Slaughtering Supervision

The adjustment of responsibility for slaughtering supervision is a crucial part of institutional reform of the State Council and transformation of government functions and is also a crucial part of the structural reform of food safety supervision and administration. In 2013, the supervision function of live pig slaughtering was formally transferred to MOA from MOC. In 2014, MOA proactively promoted the structural reform and function transference of slaughtering administration, and accelerated the function adjustment of slaughtering supervision.

6.1 Boosting function adjustment of slaughtering supervision

The nationwide function adjustment of slaughtering supervision was boosted. A series of documents were issued by MOA, including *Notice on Solving Transitional Problems after Function Adjustment of Slaughtering Supervision* (Nongyifa〔2014〕No.1) and the *Notice of the General Office of MOA on Promoting Function Adjustment of Slaughtering Supervision* (Nongbanyi〔2014〕NO.47). The national symposium for slaughtering supervision and the national conference for slaughtering supervision namely the conference for special rectification work of pig slaughtering were organized, in order to intensify inspection and comprehensively deploy and promote the function adjustment of slaughtering supervision. By the end of 2014, the function transference of live pig slaughtering supervision from the Ministry of Commerce to the Ministry of Agriculture was completed, after which the Office of Slaughtering Administration of MOA was established under the Veterinary Bureau, and Division of Slaughtering Industry Administration was set up, and the Center of Slaughtering Technology of MOA was established under China Animal Disease Control Center. The function adjustment completion rates of the provincial, municipal and county levels were 97%, 51% and 36% respectively.

6.2 Intensifying slaughtering supervision

Slaughtering supervision was intensified to guarantee meat product quality and safety in the link of slaughtering. The official veterinarian dispatch system was implemented, and the special rectification activity of live pig slaughtering was carried out. The supervision and sampling inspection for "lean meat power" in pig slaughtering was strengthened, and the urging for inspection was intensified. The law violation actions were severely penalized, including unregistered and unregulated slaughtering, the purchase and slaughtering of animals that die of diseases, injection of water and other substances into animal products and feeding animals with "lean meat power", etc. In 2014, a total of 1387 designated slaughterhouses were closed for unqualified slaughtering conditions, with 3386 cases of illegal slaughtering investigated and penalized, and more than 7.8 million samples were detected with "lean meat power".

6.3 Advancing transformation and upgrade of slaughtering industry

The *2016-2025 Development Planning Outline of live Pig Slaughtering Industry* was drafted, which put forward objective tasks and specific measures for pig slaughtering industry. Research on relevant policies was strengthened by conducting studies on the development of slaughtering industry. The laws, regulations and standards on slaughtering were promoted by launching the revision of *Pig Slaughtering Administration Regulations*, adjusting the work scope and a part of committee members of the National Committee of Standardized Technologies of Slaughtering and Processing, and clarifying the national standards and industry standards of slaughtering.

6.4 Promoting statistical monitoring, publicity and training of slaughtering industry

The statistical monitor system of pig slaughtering was established, and the *2014-2015 Statistic Report System of the Slaughtering of Pigs and Other Livestock and Poultry* was issued to notify the statistic monitoring information of pig slaughtering to relevant authorities. The purchase price of live pig, the ex-factory price of carcass meat and the monthly amount of slaughtering in large-scale slaughtering companies were released on the portal websites of government affairs. The statistic monitoring system of the slaughtering of live pig and other livestock and poultry and the national administration information system of slaughtering industry were researched and integrated. In Inner Mongolia Autonomous Region and Anhui Province, two sessions of national training courses on slaughtering supervision and management were organized, attended by more than 500 persons in charge and excellent personnel in animal health supervision authorities at provincial and municipal levels. The monitoring on slaughtering industry and the special rectification activities of slaughtering were publicized and reported by Xinhua News Agency, CCTV, Farmers' Dairy, the website of http://www.agri.cn/ and other media.

Chapter 7

Domestic and International Exchanges and Cooperation

In 2014, China continued to promote exchanges and cooperation in the veterinary field, earnestly implemented animal health international obligations, deepened bilateral and multilateral exchanges and cooperation with international organizations and other countries, strengthened exchanges and cooperation with Hong Kong, Macao and Taiwan regions, promoted healthy development of domestic veterinary cause and made due contributions to global animal health and public health security.

7.1 Exchanges and Cooperation with International Organizations

7.1.1 Deepen Exchanges and Cooperation with OIE

In 2014, our country actively performed its obligations as OIE member, and fully participated in OIE related activities. China was officially recognized by OIE as negligible bovine spongiform encephalopathy risk country and African horse sickness historically non-infected country, and 3 institutions were recognized as OIE reference centers.

7.1.1.1 Actively Perform International Obligations as OIE Member

In 2014, China notified to OIE animal disease information accurately and timely, paid membership dues in full, and participated in a series of conferences and activities held by OIE. In May 2014, MOA delegation group participated in the 82nd General Session of the World Assembly of Delegates, presided over the meeting of OIE regional commission for Asia, the Far East and Oceania, and made special report on regional activities in 2013; as the vice chairman country of executive committee of OIE Southeast Asia-China Foot-and-Mouth Disease Control Campaign (SEACFMD), attended SEACFMD chairman and vice chairman meeting, the 20th SEACFMD committee meeting, SEACFMD 17th national coordinator meeting; as chairman country of regional executive committee of Global Framework for Transboundary Animal Diseases (GF-TADs), chaired the 8th executive committee meeting of OIE Asia Pacific animal disease prevention and control framework, attended GF-TADs 7th global executive committee meeting and make regional work report.

7.1.1.2 Actively organized OIE Meetings

In April 2014, the Veterinary Bureau of MOA jointly hosted OIE Regional Information Seminar for Recently Appointed OIE Delegates (Asia) with OIE regional commission for Asia, the Far East and Oceania in Beijing, representatives from the OIE headquarter, Regional Representative for Asia and the Pacific, OIE Sub-Regional Representative for South-East Asia, the Philippines, Mongolia, India, China and more than 10 countries attended this Seminar (Figure 7-1).

Chapter 7　Domestic and International Exchanges and Cooperation

Figure 7-1　OIE Regional Information Seminar for Recently Appointed OIE Delegates (Asia) (Beijing)

In September 2014, the Veterinary Bureau of MOA and the Lanzhou Veterinary Research Institute of Chinese Academy of Agricultural Sciences sponsored Office International des Epizooties (OIE) /Japan Trust Fund (JTF) joint program "Asian Foot-and-mouth Disease Control Plan" the 3rd session of coordinating committee enlarged meeting, communicated foot-and-mouth disease prevalence in 2013-2014 and progress in prevention & control, and fully discussed foot-and-mouth disease prevention & control targets and road-map.

7.1.1.3　Follow-up and Participation in OIE Standard Revision for Animal Health

In 2014, China continued to actively participate in the international animal health standards formulation and revision work, based on the translation and publishing of "Code", "OIE PVS Tools" and other OIE publications, deliberated several times on OIE "Terrestrial Animal Health Code", "Manual of Diagnostic Tests and Vaccines for Terrestrial Animals", "Aquatic Animal Health Code", "Manual of Diagnostic Tests for Aquatic Animals" revision contents and OIE "the 6th Strategic Plan 2016-2020", and totally submitted 56 related appraisal suggestions and recommendations to OIE, among which 24 were adopted by OIE.

7.1.1.4 Actively fulfill the duties as OIE Reference Center

In May 2014, the 82nd General Session of the World Assembly of OIE Delegates passed a resolution to recognize National Diagnostic Center for Exotic Animal Diseases of China Animal Health and Epidemiology Center as OIE reference laboratory on Peste des petits ruminants; Institute of Zoonosis of Jilin University as OIE collaborating center on Food-Borne Parasites from the Asia-Pacific Region; recognize China Animal Health and Epidemiology Center as OIE Collaboration Center on Veterinary Epidemiology, to jointly undertake OIE Collaborating Centre responsibility from the Asia-Pacific Region on Veterinary Epidemiology and Public Health with Massey University of New Zealand. By the end of 2014, China has totally been recognized 12 reference laboratories and 3 collaborating centers by OIE.

In 2014, relevant laboratories actively performed their duties as OIE reference laboratories and collaborating centers, earnestly participated in or organized OIE related conferences and training workshops, such as participating in "The 3rd Session of OIE Global Reference Center Conference", "The 20th Meeting of OIE Southeast Asia-China Foot-and-Mouth Disease Control Campaign(SEACFMD) ", "The 9th OIE/FAO Foot-and-mouth Disease Reference Laboratories Network Meeting", "The 13th Asia-Pacific OIE Aquatic Animal Health Conference", " OIE National Focal Points for Animal Welfare Enlarged Meeting " etc., organized "Asian Swine Disease Prevention and Control Project Workshop", "The 3rd International Symposium on Newcastle Disease and Peste des Petits Ruminants", "OIE Southeast Asia Rabies Diagnosis Training workshop", etc. (Figure 7-2 to Figure 7-4), provided standard substances and test reagents to OIE members, participated in laboratory test capacity comparison in relevant region, completed testing and confirmation of related OIE member sample submission, and well performed duties as OIE reference center (laboratory).

Figure 7-2　Asian Swine Disease Prevention & control Project Workshop (Beijing)

Figure 7-3　The 3rd International Symposium on Newcastle Disease and Peste des Petits Ruminants (Qingdao)

Figure 7-4　OIE Southeast Asia Rabies Diagnosis Workshop (Changchun)

7.1.1.5 Participation in OIE Animal Welfare Fund related Work

In 2014, China donated additional $600,000 to OIE Animal Health and Welfare Fund, with cumulative contribution of $800,000 in funding and launching OIE Asian Pig Disease Prevention & control Project, to improve pig disease prevention &control and diagnosis ability; attended OIE Animal Health and Welfare Fund committee meeting, assigned a specialist to visit OIE Regional Representative for Asia and the Pacific, deeply involved in OIE related projects, and promoted China's cooperation with OIE. In addition, actively followed up OIE animal welfare related work, in November 2014 assigned personnel to attend OIE National Focal Points for Animal Welfare Enlarged Meeting held in Canberra, the capital of Australian (Figure 7-5), and reported on "China's Practice Experiences and Challenges on OIE Animal Welfare Standards".

Figure 7-5　OIE National Focal Points for Animal Welfare Enlarged Meeting (Canberra, Australia)

7.1.2　Strengthen Exchanges and Cooperation with FAO

In 2014, China continued to strengthen exchanges and cooperation with FAO in the field of animal health, held the 2nd session of animal health cooperation consultation conference, continued to implement Field Epidemiology Training Program for Veterinarians (FETPV), and conducted African Swine Fever Risk Prevention Project, etc. to constantly improve China's animal disease prevention & control ability.

7.1.2.1 The 2nd Session of Animal Health Cooperation Consultation Conference

On May 22nd to 23rd, 2014, the Veterinary Bureau of MOA held the 2nd Session of China-FAO Animal Health Cooperation Consultation Conference with FAO Animal Production and Health Division at FAO headquarter in Rome (Figure 7-6), exchanged basic situations of China and FAO in the field of animal health, discussed H7N9 influenza and other public health emergencies, as well as peste des petits ruminants and other trans-boundary animal disease prevention & control hot issues, clarified future veterinary cooperation direction and strategies, reached a consensus to strengthen cooperation in trans-boundary animal disease prevention & control, veterinary laboratory network and capacity building, peste des petits ruminants and other high-risk exotic animal diseases prevention, and veterinary experts resources sharing, etc.

Figure 7-6 The 2nd Session of Animal Health Cooperation Consultation Conference (Rome, Italy)

7.1.2.2 Continuous Cooperation with FAO on Field Epidemiology Training Program for Veterinarians

In July 2014, western region veterinary epidemiology basic training and veterinary epidemiology advanced training workshop and the 3rd project implementation seminar were held respectively (Figure 7-7), totally more than 100 people attended the training; In September 2014, the 2nd session of China FETPV was completed smoothly, 19 students from national and provincial animal health institutions successfully completed veterinary epidemiology courses after two years of systematic study, all of whom graduated as scheduled (Figure 7-8). In November 2014, the 3rd session of China FETPV basic training class began and 40 epidemiological key technical personnel attended the training.

Figure 7-7 Veterinary Epidemiology Advanced Training Workshop and The 3rd Session of Project Implementation Seminar (Beijing)

Figure 7-8 The 2nd Session of China Veterinary Field Epidemiology Training Project Graduation Ceremony (Qingdao)

Chapter 7　Domestic and International Exchanges and Cooperation

7.1.2.3　Joint Implementation of "China-Africa Swine Fever Prevention Project"

In July and November 2014, MOA and FAO Representative Office in Beijing jointly organized "China-Africa Swine Fever Prevention & control Project (ASF–TCP) Kick-off Meeting" and "Africa Swine Fever Prevention and Control Policy Consultation Conference", sharing experiences, analyzing situations, exchanging control policy and measures, to strengthen building of joint defense and control mechanism, research and reserve for laboratory diagnosis technology, improve comprehensive prevention & control ability, carry out joint research to promote China's African swine fever prevention, to prevent African swine fever from cross-border transmission.

7.1.2.4　Joint Implementation of "Strengthening Bio-Safety and Carrying Out Market Value Chain Analysis and Cooperation Project"

In order to promote China's live poultry market bio-safety level and disease spread risk prevention capacity, Veterinary Bureau of MOA, together with FAO in three provinces including Hunan, Yunnan and Guangxi, carried out "Regarding Helping Guangxi, Hunan, and Yunnan Provinces Strengthen Bio-Safety and Carry Out Market Value Chain Analysis Cooperation Project" (OSRO/GLO/302/USA). In September 2014, project symposium was held in Changsha city, Hunan province, discussing and exchanging on bio-safety reformation of live poultry wholesale market, live poultry market value chain risk assessment, etc, and making the value chain analysis and investigation plan for the live poultry market. In December 2014, live poultry market value chain evaluation and bio-safety seminar was held in Kunming city, Yunnan province, discussing improvement of live poultry market bio- safety, poultry value chain structure and disease risk analysis and other contents.

7.1.3　Exchange and Cooperation with the World Bank

Execute World Bank Grants China Emerging Infectious Diseases Project (Phase 3). In January 2014, International Conference on Brucellosis Prevention and Control Strategy and Completion Workshop of China Emerging Infectious Diseases Project by the World Bank was successfully held in Beijing (Figure 7-9). Experts made 28 theme reports on brucellosis epidemiological researches and coping strategies, diagnosis and laboratory research, prevention & control vaccine research, report and monitoring, facilitating behavior change and other five topics, carried out experts focus discussion, to form national brucellosis prevention & control strategic policy proposal (draft) and passed the deliberation at the conference. By the chance of the World Bank emerging infectious disease project, five veterinary personnel was selected to participate in epidemiology training at Massey University of New Zealand, who have completed their studies and got the master's degree in veterinary medicine.

Figure 7-9　International Conference on Brucellosis Prevention and Control Strategy and Completion Workshop of China EID Project (Beijing)

7.2 Bilateral and Multilateral Exchanges and Cooperation

In 2014, China continued to strengthen bilateral and multilateral exchanges and cooperation in veterinary field, signed relevant cooperation agreements, enhanced information communication, organized or participated in seminars, workshops. etc, to continually improve animal disease joint prevention & control capability.

7.2.1 Exchanges and Cooperation with Neighboring Countries

From Aug 20th to 22nd 2014, with funding of FAO and Asian Development Bank, veterinary departments of three governments of China, Mongolia and Russia held a seminar on trans-boundary animal disease prevention & control in Erguna city, Inner Mongolia. The three parties of China, Mongolia and Russia analyzed epidemic situation of foot-and-mouth disease, avian flu, brucellosis, and African swine fever, and had exchanges on technical measures of prevention & control. Consensus was reached that "One World, One Health" concept should be effectively established, to break through boundaries of countries, departments and disciplines, integrate and leverage all resources, strengthen bilateral and multilateral veterinary cooperation, effectively cope with challenges of constantly emerging animal disease risk, to ensure public health safety; in trans-boundary animal disease prevention & control, establish close joint prevention & control cooperation mechanism, reinforce animal disease diagnosis technology and comprehensive prevention & control measures of cooperative research, encourage scientific research units and veterinary drugs and veterinary biological products manufacturing enterprises to carry out exchanges and cooperation each other. During the meeting, the Veterinary Bureau of MOA held bilateral talks with Mongolia and Russia respectively, and carried out in-depth communication on joint surveillance at cross-border regions and implementation mutual recognition of zones free from specified animal diseases and compartment, facilitating animals and animal products trade, etc.

On October 16th 2014, China-Laos Trans-boundary Animal Disease Prevention and Control Project Feasibility Study Meeting Minutes Signing Ceremony was held at the Animal Husbandry and Fishery Department of Laos Ministry of Agriculture and Forestry. The Aid to Laos Trans-boundary Animal Disease Prevention and Control Program is a foreign aid project to promote cooperation between China and Laos in respect of trans-boundary animal disease prevention and control, and the project main sites were identified in Vientiane city, Louang Namtha province and Phôngsali province of Laos, among which trans-boundary animal disease testing and Diagnosis Laboratory would be built in Vientiane city, and a trans-boundary animal disease surveillance station would be built in Louang Namtha and Phôngsali province respectively.

On November 14th 2014, Han Changfu, the Minister of Agriculture, and UOhn Myint, Burma's Minister of Animal Husbandry and Aquaculture and Rural Development jointly signed the "Sino-Burmese Animal Husbandry and Fishery Cooperation Memorandum of Understanding", in the next five years China would train 300 agricultural technology and management personnel for Burma, set up Sino-Burma agricultural technology training center in Yunnan, and construct the agricultural technology demonstration center and region free from specified diseases in Burma, offer small agricultural loans, to promote in-depth development of Sino-Burma agricultural cooperation.

On September 1st 2014, Sino-Ukrainian cooperation committee agricultural cooperation sub-committee the 4th meeting was held in

Ukraine. The meeting summed up all the work carried out since the 3rd meeting, explored the possibilities to expand cooperation in fishery freshwater aquaculture, fungus production, veterinarian and veterinary drug and other aspects. After the meeting, "Agreement of the Government of P. R. China and the Government of Ukraine on Cooperation in the Field of Animal Health and Quarantine" was reached and "the 4th Meeting Minutes of Sino-Ukraine Cooperation Committee Agricultural Cooperation Sub-committee" was signed.

7.2.2 Exchanges and Cooperation between China and European Union

Continue to implement the EU's Framework Program 7 (FP7) Sino-Europe trans-boundary animal disease and zoonosis epidemiological and laboratory research (Link TADs) project. In April 2014, FP7-Link TADs project seminar was held in Shanghai, carrying out discussions surrounding H5N1, H7N9 influenza, African swine fever, foot-and-mouth disease, brucellosis and other major animal diseases and zoonosis epidemiology and prevention and control technology. In June 2014, China and Denmark jointly organized the 1st Sino-Denmark Veterinary Work Meeting, exchanging issues such as animal disease prevention& control, drug resistance risk management and other situations of both parties, studying the direction of cooperation of the next step, and proposed to strengthen communication and coordination in animal products quality safety management and traceability, major animal disease purification, prevention & control and other key areas as well as in OIE and CAC and other related international veterinary affairs. In addition, China Animal Health and Epidemiology Center signed memorandum of cooperation with German Federal Animal Health Research Institute, further deepening the bilateral cooperation in veterinary epidemiology and risk assessment, etc.

7.2.3 Exchanges and Cooperation with Other Countries

In June 2014, the Vice Minister of Argentina Animal Husbandry and Fishery visited China, and exchanged views on strengthening Sino- Argentina agricultural cooperation. In July 2014, Han Changfu, the Minister of MOA visited Argentina, and signed" Cooperation Memorandum of Understanding between MOA of the P. R. China and the Ministry of Animal Husbandry and Fisheries of the Republic of Argentina on Veterinary Health Cooperation", and came into new cooperation agreement on the next step of Argentina beef import.

7.3 Exchanges and Cooperation with Hong Kong, Macao and Taiwan regions of China

7.3.1 Hong Kong and Macao

The Veterinary Bureau of MOA signed "Veterinary Cooperation Arrangement of MOA of P. R. China and the Food and Health Bureau of the Government of Hong Kong Special Administrative Region" and "Veterinary Cooperation Arrangement of MOA of the P. R. China and the Civil Affairs Department of the Government of Macao Special Administrative Region", continued to deepen exchanges and cooperation in veterinary legislation, licensed veterinarian management, information exchange, major animal diseases and key zoonosis prevention & control, building and management of zone free from animal diseases, international veterinary affairs, personnel training, etc (Figure 7-10). In the application of free from AHS recognition, full consideration was given to requirements of Hong Kong and Macao, active coordination was made with OIE headquarter, while issuing the certificate of China free from African horse sickness; meanwhile, use OIE

Director-general signed letter to prove that Hong Kong and Macao had no epidemic existence. Provide hemagglutination inhibition test positive serum of avian influenza virus sub-type H7 (H7N9 strain) to Agriculture, Fisheries and Conservation Department of Hong Kong. Provide avian flu prevention & control measures, equine influenza and other horse disease technology and animal rabies laboratory diagnosis technical training to Macao.

Figure 7-10 MOA Signed Veterinary Cooperation Arrangements with Hong Kong and Macao Respectively (Beijing)

7.3.2 Taiwan

Make in-depth communication and discussion with Taiwan delegation on exchanges and cooperation in veterinary field across Taiwan straits, H7N9 influenza prevention & control, migration of migratory birds and early warning and surveillance of avian flu etc.; assign personnel to attend 2014 Veterinary Management and Technical Seminar Across Taiwan Straits, to exchange situations on licenesed veterinarians, animal medical institutions, animal disease prevention system, veterinary drug registration and major animal disease emergency management; provide H5N2 sub-type of avian influenza virus nucleic acid sequence to Taiwan, highly pathogenic avian influenza treatment personnel protective measures and rabies prevention & control measures, oral vaccine and personnel protection measures, etc.

Annex 1
National Veterinary Laboratory Status

To improve animal disease prevention and control and technical support ability on the supervision and regulation of animal origin food safety, MOA approved 3 national veterinary reference laboratories, 4 national veterinary diagnosis Laboratories, 4 national veterinary drug residue benchmark laboratories; the Ministry of Science and Technology approved the building of 3 national key laboratories in veterinary field; MOA confirmed 2 comprehensive veterinary key laboratories, 15 professional/regional key laboratories in veterinary field, 12 agricultural scientific Observation Experiment Stations; totally 15 reference centers approved by OIE.

1. National Veterinary Reference Laboratory

According to the requirements of "*National Veterinary Reference Laboratory Management Method*", by the end of 2014, MOA approved 3 national veterinary reference laboratories, which were National Avian Flu Reference Laboratory, National Foot-and-mouth Disease Reference Laboratory and National Bovine Spongiform Encephalopathy Reference Laboratory respectively (Annex Table 1-1).

Annex Table 1-1 National Veterinary Reference Laboratories

Laboratory Name	Main Responsibilities	Institutes of Laboratories
National Avian Flu Reference Laboratory	Undertake basic researches on avian influenza, foot-and-mouth disease, bovine spongiform encephalopathy and other correlative diseases, diagnosis technology research and development, standardization of diagnostic reagents, disease confirmation, technical training respectively	Harbin Veterinary Research Institute of Chinese Academy of Agricultural Science (China Animal Health and Epidemiology Center Harbin Branch)
National Foot-and-mouth Disease Reference Laboratory		Lanzhou Veterinary Research Institute of Chinese Academy of Agricultural Sciences (China Animal Health and Epidemiology Center Lanzhou Branch)
National Bovine Spongiform Encephalopathy Reference Laboratory		China Animal Health and Epidemiology Center

2. National Veterinary Diagnosis Laboratory

By the end of 2014, MOA had totally specified 4 national veterinary diagnosis laboratories, which were Newcastle disease, classical swine fever, rinderpest and contagious bovine pleuropneumonia diagnosis laboratory (Annex Table 1-2).

Annex Table 1-2 National Veterinary Diagnosis Laboratory

Laboratory Name	Main Responsibilities	Institute of the Laboratory
National Newcastle Disease Diagnosis Laboratory	Undertake relevant basic researches on diseases, diagnosis technology research and development, standardization of diagnostic reagents, disease confirmation, technical training, etc	China Animal Health and Epidemiology Center
National Classical Swine Fever Diagnosis Laboratory		China Institute of Veterinary Drug Control
National Rinderpest Diagnosis Laboratory		China Institute of Veterinary Drug Control
National Contagious Bovine Pleuropneumonia Diagnosis Laboratory		Harbin Veterinary Research Institute of Chinese Academy of Agricultural Sciences (China Animal Health and Epidemiology Center Harbin Branch)

3. National Veterinary Drug Residue Benchmark Laboratory

By the end of 2014, based on technical superiority of relevant units, MOA has constructed 4 national veterinary drug residue benchmark laboratories (Annex Table 1-3).

Annex Table 1-3 National Veterinary Drug Residue Benchmark Laboratories

Supporting Institution	Drug Testing Scope
China Institute of Veterinary Drug Control	Fluoroquinolones, tetracycline, β-receptor agonist drugs
Animal Medical School of China Agricultural University	Abamectin, sulfonamides, nitroimidazole, chloramphenicol and zeranol drugs
South China Agricultural University	Organophosphorus, pyrethroid, β-lactam, arsthinol and diethylstilbestrol drug
Animal Husbandry and Veterinary School of Huazhong Agricultural University	Quinoline, nitrofurans, benzimidazole drug

4. National Veterinary Key Laboratory

According to the requirements of *"Management Measures for Building and Operation of National Key Laboratories"*, by the end of 2014, the Ministry of Science and Technology has approved the building of 3 national key laboratories in veterinary field, which were national key laboratory of veterinary biotechnology, livestock Etiological biology and Biological Safety of Pathogenic Microorganisms respectively (Annex Table 1-4).

Annex Table 1-4　National Veterinary Key Laboratory

Laboratory Name	Main Responsibilities	Institute of the Laboratory
National Key Laboratory of Veterinary Biotechnology	Carry out animal pathogenic gene engineering and cell engineering research, and at the same time undertake the veterinary basic theory research on molecular biology	Harbin Veterinary Research Institute of Chinese Academy of Agricultural Sciences (China Animal Health and Epidemiology Center Harbin Branch)
National Key Laboratory of Livestock Etiological Biology	With major animal disease of livestock as the research object, for the important scientific issues and key technologies on viruses, bacteria, parasites disease prevention and control, carry out research on etiology and the laws of the interaction between pathogen, host and the environment	Lanzhou Veterinary Research Institute of Chinese Academy of Agricultural Sciences (China Animal Health and Epidemiology Center Lanzhou Branch)
National Key Laboratory of Biological Safety of Pathogenic Microorganisms	With the biological safety of pathogenic microorganism as the research direction, focus on the discovery, early warning, detection and prevention related theory and technology research of pathogenic microorganisms, including: pathogenic microorganisms reconnaissance and early warning technology research, rapid test and identification technology research of pathogenic microorganisms, discovery of new infectious diseases and tracking technology research, important pathogenic microorganisms pathogenic mechanism and basic research in the prevention and control, etc.	Microbial Epidemic Research Institute of the Chinese People's Liberation Army Military Academy of Medical Sciences and Biological Engineering Research Institute

5. MOA Veterinary Key Laboratory

According to "*MOA Key Laboratory Development Plan (2010-2015)*" and "*MOA Key Laboratory Management Method*", from 2010 to 2011, MOA organized and carried out two batches of key laboratory system deployment and selection. In 2011, MOA confirmed 30 subject groups including comprehensive key laboratory, professional (regional) key laboratory and agricultural scientific observation experiment stations. Among them, in the veterinary field there are mainly 2 Subject groups including veterinary drug and veterinary biological technology group, and animal disease pathogen biology Subject group (Annex Table 1-5). Due to the existence of crossing in the key laboratory and agricultural scientific observation experiment station of two Subject groups, combined with agricultural products quality safety Subject group, which also included MOA veterinary drug residue and banned additives detection key laboratory (China Agricultural University) and MOA veterinary drug residue detection key laboratory (Huazhong Agricultural University) and other 2 key laboratories of veterinary drug residue detection, by the end of 2014, there were totally 2 national MOA comprehensive veterinary key laboratories, 15 professional/regional veterinary key laboratories, and 12 agricultural scientific observation experiment stations.

Annex Table 1-5 Veterinary Drug and Veterinary Biological Technology Subject Group

Subject Group	Category	Name	Supporting Institution
Veterinary Drug and Veterinary Biological Technology Subject Group	Comprehensive Key Laboratory (1)	MOA Veterinary Drug and Veterinary Biological Technology Key Laboratory	Harbin Veterinary Research Institute of Chinese Academy of Agricultural Sciences
	Professional/ Regional Key Laboratory (8)	MOA Veterinary Drug Research and Development Key Laboratory	Lanzhou Animal Husbandry and Veterinary Drug Research Institute of Chinese Academy of Agricultural Sciences
		MOA Veterinary Vaccines Research and Development Key Laboratory	South China Agricultural University
		MOA Veterinary Diagnostics Preparation R&D Key Laboratory	Huazhong Agricultural University
		MOA Veterinary Biological Products Engineering Technology Key Laboratory	Jiangsu Province Academy of Agricultural Sciences

Annex 1 National Veterinary Laboratory Status

(continued)

Subject Group	Category	Name	Supporting Institution
Veterinary Drug and Veterinary Biological Technology Subject Group	Professional/ Regional Key Laboratory (8)	The Ministry of Agriculture Special Animal Biologics Preparation R&D Key Laboratory	Military Academy of Medical Sciences Veterinary Institute
		MOA Fishery Drug Research and Development Key Laboratory	Pearl River Fisheries Research Institute of Chinese Academy of Fishery Sciences
		MOA Poultry Biological Preparation Research and Development Key Laboratory	Yangzhou University
		MOA Animal Immunology Key Laboratory	Henan Academy of Agricultural Sciences
	Agricultural Scientific Observation Experiment Station (10)	MOA Veterinary Drug and Veterinary Biotechnology Beijing Scientific Observation Experiment Station	Beijing Animal Husbandry and Veterinary Institute of Chinese Academy of Agricultural Sciences
		MOA Veterinary Drug and Veterinary Biotechnology Tianjin Scientific Observation Experiment Station	Tianjin Institute of Animal Husbandry and Veterinary Research
		MOA Veterinary Drug and Veterinary Biotechnology Xinjiang Scientific Observation Experiment Station	Xinjiang Uygur Autonomous Region Academy of Animal Husbandry
		MOA Veterinary Drug and Veterinary Biotechnology Shaanxi Scientific Observation Experiment Station	Northwest Agriculture and Forestry University of Science and Technology
		MOA Veterinary Drug and Veterinary Biological Technology Hubei Scientific Observation Experiment Station	Institute of Animal Husbandry and Veterinary of Hubei Academy of Agricultural Sciences
		MOA Veterinary Drug and Veterinary Biotechnology Sichuan Scientific Observation Experiment Station	Sichuan Agricultural University
		MOA Veterinary Drug and Veterinary Biotechnology Guangxi Scientific Observation Experiment Station	Guangxi Zhuang Autonomous Region Veterinary Institute
		MOA Veterinary Drug and Veterinary Biotechnology Guangdong Scientific Observation Experiment Station	Veterinary Institute of Guangdong Province Academy of Agricultural Sciences

(continued)

Subject Group	Category	Name	Supporting Institution
Veterinary Drug and Veterinary Biological Technology Subject Group	Agricultural Scientific Observation Experiment Station (10)	MOA Animal Disease Pathogen Biology East China Scientific Observation Experiment Station	Shandong Agricultural University
		MOA Animal Disease Pathogen Biology Northeast Scientific Observation Experiment Station	Northeast Agricultural University
Animal Disease Pathogen Biology Subject Group	Comprehensive Key Laboratory (1)	MOA Animal Disease Pathogen Biology Key Laboratory	Lanzhou Veterinary Research Institute of Chinese Academy of Agricultural Sciences
	Professional/Regional Key Laboratory (7)	MOA Animal Virology Key Laboratory	Zhejiang University
		MOA Animal Bacteriology Key Laboratory	Nanjing Agricultural University
		MOA Animal Parasitological Key Laboratory	Shanghai Veterinary Research Institute of Chinese Academy of Agricultural Sciences
		MOA Animal Immunology Key Laboratory	Henan Academy of Agricultural Sciences
		MOA Animal Epidemiology and Human and Animal Disease Key Laboratory	China Agricultural University
		MOA Animal Disease Clinical Diagnosis and Treatment Technology Key Laboratory	Inner Mongolia Agricultural University
		MOA Veterinary Diagnostics Preparation Research and Development Key Laboratory	Huazhong Agricultural University
	Agricultural Scientific Observation Experiment Station (6)	MOA Animal Disease Pathogen Biology North China Observation Experiment Station	Hebei Agricultural University
		MOA Animal Disease Pathogen Biology Northeast Scientific Observation Experiment Station	Northeast Agricultural University
		MOA Animal Disease Pathogen Biology East China Scientific Observation Experiment Station	Shandong Agricultural University
		MOA Animal Disease Pathogenic Biology Southwest Scientific Observation Experiment Station	Yunnan Province Animal Husbandry and Veterinary Academy of Sciences

(continued)

Subject Group	Category	Name	Supporting Institution
Animal Disease Pathogen Biology Subject Group	Agricultural Scientific Observation Experiment Station (6)	MOA Veterinary Drug and Veterinary Biotechnology Shaanxi Scientific Observation Experiment Station	Northwest Agriculture and Forestry University of Science and Technology
		MOA Veterinary Drug and Veterinary Biotechnology Xinjiang Scientific Observation Experiment Station	Xinjiang Uygur Autonomous Region Academy of Animal Husbandry

6. OIE Reference Center

By the end of 2014, China has already been recognized 15 OIE reference centers, including 12 reference laboratories and 3 collaborating centers located in 9 institutions and units (Annex Table 1-6). Moreover, Animal Influenza Laboratory in Harbin Veterinary Research Institute of Chinese Academy of Agricultural Sciences was also recognized as FAO animal influenza reference center.

Annex Table 1-6 OIE Reference Laboratories and Collaborating Centers in China

No.	Epidemic Areas	Supporting Institutions	Year of Approval
1	Avian Influenza Reference Laboratory	Harbin Veterinary Research Institute of Chinese Academy of Agricultural Sciences	2008
2	Foot-and-mouth Disease Reference Laboratory	Lanzhou Veterinary Research Institute of Chinese Academy of Agricultural Sciences	
3	Equine Infectious Anemia Reference Laboratory	Harbin Veterinary Research Institute of Chinese Academy of Agricultural Sciences	
4	White Spot Disease Reference Laboratory	Yellow Sea Fisheries Research Institute of Chinese Academy of Fishery Sciences	2011
5	Infectious hypodermal and Hematopoietic Necrosis Reference Laboratory	Yellow Sea Fisheries Research Institute of Chinese Academy of Fishery Sciences	
6	Spring Viraemia of Carp Reference Laboratory	Shenzhen Entry-exit Inspection and Quarantine Bureau	

(continued)

No.	Epidemic Areas	Supporting Institutions	Year of Approval
7	Porcine Reproductive and Respiratory Syndrome Reference Laboratory	China Animal Disease control Center	2012
8	Newcastle Disease Reference Laboratory	China Animal Health and Epidemiology Center	
9	Rabies Reference Laboratory	Changchun Veterinary Research Institute of Chinese Academy of Agricultural Sciences	
10	Collaborating Center of Zoonosis in Asia Pacific	Harbin Veterinary Research Institute of Chinese Academy of Agricultural Sciences	
11	Ovine theileriasis Laboratory	Lanzhou Veterinary Research Institute of Chinese Academy of Agricultural Sciences	2013
12	Swine Streptococcal Diseases Reference Laboratory	Nanjing Agricultural University	
13	Peste des Petits Ruminants Reference Laboratory	China Animal Health and Epidemiology Center	2014
14	Veterinary Public Health and Epidemiology Collaborating Center	China Animal Health and Epidemiology Center	
15	Asia Pacific Region Foodborne Parasitic Disease Collaboration Center	Institute of Animal and Human Diseases of Jilin University	

Annex 2
Colleges and Universities with Veterinary Major

Annex Table 2-1 shows the colleges and universities with veterinary major.

Annex Table 2-1 Information Table of Colleges and Universities with Veterinary Major

Colleges and Universities	Attributes and Characteristics of Colleges and Universities	Master and Doctor's Degree Authorization
China Agricultural University	Key university directly under the Ministry of Education, and specified in 1954 by the central government as one of 6 key universities.	Doctor's Degree of First Grade Discipline Authorization
Nanjing Agricultural University	Key university directly under the Ministry of Education, first grade discipline of national key discipline.	Doctor's Degree of First Grade Discipline Authorization
Huazhong Agricultural University	Key university directly under the Ministry of Education	Doctor's Degree of First Grade Discipline Authorization
Jilin University	Key university directly under the Ministry of Education, second grade discipline of national key discipline.	Doctor's Degree of First Grade Discipline Authorization
Northeast Agricultural University	Second grade discipline of national key discipline	Doctor's Degree of First Grade Discipline Authorization

(continued)

Colleges and Universities	Attributes and Characteristics of Colleges and Universities	Master and Doctor's Degree Authorization
South China Agricultural University	Second grade discipline of national key discipline	Doctor's Degree of First Grade Discipline Authorization
Yangzhou University	Provincial joint building university, provincial key university, with excellent engineer training program and outstanding agriculture&forestry talents education program, second grade discipline of national key discipline	Doctor's Degree of First Grade Discipline Authorization
Northwest Agriculture and Forestry University	Key university directly under the Ministry of Education, second grade discipline of national key discipline	Doctor's Degree of First Grade Discipline Authorization
Gansu Agricultural University	Provincial joint building university, provincial key university	Doctor's Degree of First Grade Discipline Authorization
Inner Mongolia Agricultural University	State Forestry Administration and the Inner Mongolia Autonomous Region joint building key university, Midwest colleges and universities fundamental ability building agricultural university	Doctor's Degree of First Grade Discipline Authorization
Guangxi University	Provincial joint building university, Midwest colleges & universities West China Comprehensive Improvement Project University	Doctor's Degree of First Grade Discipline Authorization
Sichuan Agricultural University	University with the feature of biological science and technology, with advantages in agricultural science and technology	Doctor's Degree of First Grade Discipline Authorization
Shandong Agricultural University	Provincial key university, Shandong Featured University Project, provincial joint building university	Doctor's Degree of First Grade Discipline Authorization
Shanxi Agricultural University	Provincial joint building university	Doctor's Degree of First Grade Discipline Authorization
Henan Agricultural University	Provincial joint building university, 2011 plan leader university, characteristic key discipline project building university	Doctor's Degree of First Grade Discipline Authorization
Heilongjiang Bayi Agricultural University	Provincial full-time ordinary university	Doctor's Degree of First Grade Discipline Authorization

(continued)

Colleges and Universities	Attributes and Characteristics of Colleges and Universities	Master and Doctor's Degree Authorization
Hunan Agricultural University	"Midwestern Colleges and Universities Fundamental Ability Building Engineering" university, provincial joint building university	Doctor's Degree of Second Grade Discipline Authorization
Jilin Agricultural University	Provincial key university	Doctor's Degree of Second Grade Discipline Authorization
Zhejiang University	University directly under the Ministry of Education, provincial joint building and joint administration, C9 league.	Doctor's Degree of First Grade Discipline Authorization
Guizhou University	Provincial joint building university, one of the colleges and universities of national "Midwest Colleges and Universities Comprehensive Strength Enhancement Project"	Master's Degree of First Grade Discipline Authorization
Fujian Agriculture and Forestry University	Provincial joint building university, key building university of Fujian province	Master's Degree of First Grade Discipline Authorization
Southwest University	Provincial joint building university, key university directly under the Ministry of Education.	Master's Degree of First Grade Discipline Authorization
Xinjiang Agricultural University	Provincial key university, the State Forestry Administration and joint building university in Xinjiang Uygur Autonomous Region	Master's Degree of First Grade Discipline Authorization
Anhui Agricultural University	Provincial joint building university, provincial key university, the Midwest university fundamental capacity building project	Master's Degree of First Grade Discipline Authorization
Yunnan Agricultural University	Provincial key university of Yunnan province	Master's Degree of First Grade Discipline Authorization
Hebei Agricultural University	Provincial joint building university, selected into "Midwest Colleges and Universities Fundamental Capacity Building Project"	Master's Degree of First Grade Discipline Authorization
Jiangxi Agricultural University	National key university, selected into "Midwest Colleges and Universities Fundamental Ability Building Project" university, provincial joint building university	Master's Degree of First Grade Discipline Authorization
Qingdao Agricultural University	Provincial key building university, Shandong characteristic famous university project	Master's Degree of First Grade Discipline Authorization

(continued)

Colleges and Universities	Attributes and Characteristics of Colleges and Universities	Master and Doctor's Degree Authorization
Beijing Agricultural College	Beijing agriculture and forestry university	Master's Degree of First Grade Discipline Authorization
Shenyang Agricultural University	"Midwest Colleges and Universities Fundamental Ability Building Project" key building university	Non
Tianjin Agricultural College	Tianjin ordinary undergraduate college and university	Master's Degree of First Grade Discipline Authorization
Shihezi University	One of the colleges and universities in the national "Midwest Colleges & Universities Comprehensive Strength Enhancement Project", "Midwest Colleges and Universities Fundamental Ability Building Project" key building university, the Ministry of education and Xinjiang joint building university	Master's Degree of First Grade Discipline Authorization
Northeast Forestry University	Key university directly under the Ministry of Education.	Non
Yanbian University	"Midwest Colleges and Universities Fundamental Ability Building Project" key building university	Master's Degree of First Grade Discipline Authorization
Henan University of Science and Technology	Provincial key university	Master's Degree of First Grade Discipline Authorization
Ningxia University	Ningxia Hui Autonomous Region People's Government and the Ministry of Education joint building comprehensive university, one of the universities in the national "Midwest Colleges & Universities Comprehensive Strength Enhancement Project"	Master's Degree of First Grade Discipline Authorization
Henan Institute of Science and Technology	Provincial ordinary undergraduate college and university	Master's Degree of First Grade Discipline Authorization
Zhejiang Agriculture and Forestry University	Provincial full-time undergraduate university	Non
Southwest University for Nationalities	University directly under the Ministry of Education.	Master's Degree of First Grade Discipline Authorization
Yangtze University	Provincial joint building university	Non
Northwest University for Nationalities	University directly under the Ministry of Education	Master's Degree of First Grade Discipline Authorization

Annex 2 Colleges and Universities with Veterinary Major

(continued)

Colleges and Universities	Attributes and Characteristics of Colleges and Universities	Master and Doctor's Degree Authorization
Tibet University	Joint building university of the people's government of Tibet Autonomous Region and the Ministry of Education	Master's Degree of Second Grade Discipline Authorization
Foshan Institute of Science and Technology	University directly under the Ministry of Education	Master's Degree of First Grade Discipline Authorization
Hebei North University	University directly under the Ministry of Education	Master's Degree of First Grade Discipline Authorization
Inner Mongolia University for Nationalities	Joint building provincial key university of State Ethnic Affairs and the Inner Mongolia Autonomous Region	Master's Degree of First Grade Discipline Authorization
Tarim University	Provincial joint building university	Master's Degree of First Grade Discipline Authorization
Anhui Science and Technology College	Provincial undergraduate college and university	Non
Anyang Institute of Technology	Provincial and municipal joint building undergraduate college and university	Non
Guangdong Ocean University	Joint building university of State Oceanic Administration, and the people's government of Guangdong Province	Non
Hainan University	Joint building university of the people's government of Hainan province, the Ministry of Education and Ministry of Finance	Non
Hebei University of Engineering	Provincial joint building university	Non
Hebei Normal University of Science & Technology	Provincial joint building university.	Non
Henan Institute of Animal Husbandry Economy	Provincial college and university	Non
Heze College	Provincial college and university	Non
Changchun Institute of Technology	Provincial ordinary college and university	Non
Jilin Institute of Agricultural Science and Technology	Provincial ordinary college and university	Non

(continued)

Colleges and Universities	Attributes and Characteristics of Colleges and Universities	Master and Doctor's Degree Authorization
Jinling Institute of Technology	Ordinary undergraduate college and university	Non
Liaoning Medical University	Provincial ordinary university	Non
Liaocheng University	Provincial ordinary university	Non
Linyi University	Provincial ordinary university	Non
Longyan College	Provincial and municipal joint building college and university	Non
Qinghai University	Provincial joint building university	Non
Sichuan Institute for Nationalities	Provincial college and university	Non
Xichang College	Provincial college and university	Non
Xinyang College of Agriculture and Forestry	Provincial college and university	Non
Yichun College	Provincial college and university	Non
Liaodong College	Provincial college and university	Non
Shenyang Institute of Technology	Provincial college and university	Non